Exploring Earth
from Space

Other Books in the
DISCOVERING EARTH SCIENCE SERIES

VOLCANOES AND EARTHQUAKES (No. 2842)

The first book in the series, this volume concentrates on the geologic phenomenon of Earth, and how this phenomenon has affected life on our planet. Also included is a history of the planet's geology.

VIOLENT STORMS (No. 2942)

This second book in the series deals with the atmospheric and climatic phenomena of Earth, as well as their effect on man and their influence on our planet.

MYSTERIOUS OCEANS (No. 3042)

The third volume in the series looks at the hydrologic phenomenon of Earth—its origin, purpose, mechanisms, and effect on life. The book concentrates on oceans' role on our planet.

THE LIVING EARTH: THE COEVOLUTION OF THE PLANET AND LIFE (No. 3142)

The fourth book in the series focuses on the biological phenomenon of Earth. It looks at the origins, history and future of life, as well as the effects that the geologic, climatic, and hydrologic phenomena have on life.

Exploring Earth from Space

JON ERICKSON

Discovering
Earth Science

TAB BOOKS Inc.
Blue Ridge Summit, PA

FIRST EDITION
FIRST PRINTING

Copyright © 1989 by TAB BOOKS Inc.
Printed in the United States of America

Library of Congress Cataloging-in-Publication Data

Erickson, Jon, 1948-
    Exploring earth from space / by Jon Erickson.
       p.   cm.
    Bibliography: p.
    Includes index.
    ISBN 0-8306-8842-0   ISBN 0-8306-3242-5 (pbk.)
    1. Remote sensing.   I. Title.
G70.4.E75   1989   88-36713
                                    CIP

TAB BOOKS Inc. offers software for sale. For information and a catalog, please contact TAB Software Department, Blue Ridge Summit, PA 17294-0850.

Questions regarding the content of this book should be addressed to:

    Reader Inquiry Branch
    TAB BOOKS Inc.
    Blue Ridge Summit, PA 17294-0214

Cover photographs courtesy of NASA.

Suzanne L. Cheatle: Editor
Jaclyn B. Saunders: Series Design
Katherine Brown: Production

Contents

14.45

Acknowledgments

The following organizations are recognized for their valuable assistance in providing photographs for this book: the Department of Agriculture, Soil Conservation Service (SCS); the National Aeronautics and Space Administration (NASA); the National Oceanic and Atmospheric Administration (NOAA); the National Park Service; the U.S. Air Force (USAF); the U.S. Army Corps of Engineers; the U.S. Coast Guard (USCG); the U.S. Navy (USN); and the U.S. Geological Survey (USGS).

Exploring Earth from Space

Introduction

IN November 1982, Typhoon Iwa bore down on the Hawaiian Islands. The only warning the inhabitants had of its approach was from the GOES weather satellite centered over the Pacific Ocean. Meteorologists had monitored Iwa closely as it developed southwest of Hawaii. After moving northward for several days, the typhoon abruptly turned toward the islands, and within 24 hours, it struck Kauai and Oahu. Based on satellite imagery of the storm, the National Weather Service was able to give timely typhoon warnings, and only one life was lost even though property damage in the heavily populated coastal areas totaled nearly a quarter of a billion dollars.

The 1984 African drought resulted in upwards of a million people dead or dying from famine. To prevent the recurrence of such a tragedy, investigators are using satellites to map vegetation across the entire African continent. On the edges of deserts, satellite imagery chronicles just where the grassland is disappearing. Computer-enhanced imagery from Landsat satellites can determine the amount of stress that vegetation is experiencing during a drought. In addition, the satellite imagery can depict how much vegetative cover is lost from one year to the next. Satellites also might help answer the larger ecological questions about vanishing forests, soil erosion, and burgeoning human populations.

Our world is shrinking day by day as finite natural resources are dwindling to feed man's insatiable appetite for fossil fuels and minerals. The developed countries, which include only 10 percent of the world's population, consume some 90 percent of its wealth. No wonder developing countries look to the developed countries with envy and urgently desire to improve their standard of living by exploiting natural resources. However, the increased industrialization needed for such improvements will place an additional strain on the world's reserves, as well as produce more pollution, which is an unfortunate consequence of prosperity. The use of satellite imagery to map terrain features that signify oil traps or mineral deposits in unexplored regions will aid substantially in keeping up with demand. The imagery is also useful for tracking the effects of industrial pollution on the environment.

Space exploration and the use of space technology for remote sensing of the Earth is one of the most fascinating fields of earth science. The first half of this book deals with the technology and the various types of remote-sensing techniques. The book begins with the development of rocketry, the launching of space probes, and the development of earth satellites. It discusses how satellite and airborne imagery is obtained, processed, and used in the field. The second half deals with the application of this technology and the way it can help solve many of the world's serious problems. The discussion continues with the use of satellites in forecasting weather, monitoring the oceans, planning land use, performing geologic mapping, farming, and delineating potential natural disasters. Through the use of satellite technology, we can monitor the Earth on a global scale and protect and preserve it for future generations.

1

The
Discovery of Space

EVER since man first gazed upon the stars, he has wondered what awaits him in the heavens. Around 5000 years ago, in what is now Iraq, the Sumerians and their Babylonian successors charted the skies and named many constellations. The Greeks were the first true astronomers, and as early as 300 B.C., they reasoned that the Earth was round because it cast a round shadow on the Moon.

In 240 B.C., the Greek astronomer Eratosthenes estimated the size of the Earth by measuring the difference in altitude of the Sun between what is now Aswan, Egypt, and Alexandria, 500 miles to the north. The Sun at Alexandria was seven degrees lower on the horizon than in Aswan, and Eratosthenes used these figures to calculate that the Earth was therefore roughly 25,000 miles in circumference and about 8,000 miles in diameter. In the second century B.C., the Greek astronomer Hipparchus observed the Earth's shadow across the Moon and determined

from this that the distance of the Moon was thirty times the diameter of the Earth, or about 240,000 miles.

The Earth also was thought to be the center of the universe, with the Sun, the planets, and all the stars revolving around it. This misconception was not dispelled until the middle of the sixteenth century, when the Polish astronomer Nicolaus Copernicus suggested that the Earth and all the other planets revolved around the Sun and that the Moon orbited the Earth. With the invention of the telescope by Galileo in the early 1600s, the heavens came into focus for the first time. Since then, people have dreamed of machines that would take them to other celestial worlds.

THE EARLY FLIGHTS

Man's desire to leave the Earth can be seen in Greek mythology and science. In A.D. 160, the Greek satirist Lucian of Samosata wrote an account of Ulysses' ship having been caught up in a whirlwind and carried on a seven-day voyage to the Moon.

One of the first jet-propelled devices was built by Heron of Alexandria during the first century B.C. His invention, the *aeolipile*, was a hollow sphere with two nozzles on either side and mounted on pivots to a vessel filled with water (FIG. 1-1). When a fire was lit below, the water boiled, the steam blew out the nozzles, and the device spun around like a modern-day lawn sprinkler.

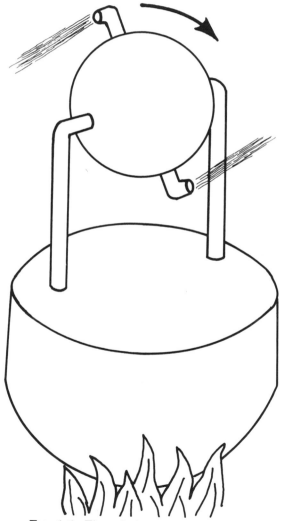

FIG. 1-1. The principle of the aeolipile.

A more practical means of propelling objects came with the invention of gunpowder in China during the middle of the eleventh century. The first rockets were reportedly used by the Chinese against invading Mongols in A.D. 1232. In 1500, a Chinese official named Wan Hu tried to fly in a rocket-powered chair using 47 large rockets ignited by 47 servants, only to vanish without a trace in a gigantic burst of flame.

The first free flight by humans was a hot-air balloon ascension from the gardens of the Chateau de la Muette outside Paris, France, on November 21, 1783. The balloon was designed by two brothers, Joseph and Jacques Montgolfier. It stayed aloft for some 25 minutes and came down in an open field about 5 miles away with its two passengers unharmed.

The Montgolfier brothers became fascinated with the idea of flight when they discovered that a paper bag filled with smoke would rise in the air. In their first public demonstration, the brothers used a large spherical sack made of linen and lined with paper. It was 36 feet across and weighed about 500 pounds. When the balloon was inflated over a hot fire, it rose to a considerable height, stayed aloft for about 10 minutes, and came down about 1½ miles away. In their next demonstration, they fitted a balloon with a small cage that carried a sheep, a rooster, and a duck. The flight ended in the woods about 2 miles away, and the travelers came down uninjured.

The French physicist Jacques Charles built the first hydrogen-filled balloon out of silk covered with rubber. Hydrogen was obtained by the reaction of sulfuric acid on iron filings. It took several days and used nearly 500 pounds of acid and 1,000 pounds of iron to inflate the leaky balloon. The 13-foot-diameter balloon was first launched on August 27, 1783, stayed aloft for 45 minutes, and traveled a distance of 15 miles. In August 1784, the French chemist Guyton de Moreau and a companion made a high-altitude balloon flight of more than 10,000 feet to collect data on the temperature and pressure of the atmosphere. Their craft was thus the world's first meteorological balloon.

In the 1890s, weather instruments were routinely carried aloft by balloons. In 1928, balloon-borne radiosonde instruments were developed to transmit meteorological data directly to the ground. One of the first to recognize the value of the hydrogen-filled balloon in warfare was Benjamin Franklin, who envisioned using thousands of balloons to transport troops into enemy territory.

Aerial observations of the battlefield were made by balloonists during the American Civil War. Needless to say, they were ideal targets and risked life and limb in order to get a glimpse of the enemy from above.

THE DREAM OF SPACE TRAVEL

It is basic to human nature to explore at first the world immediately around us and then the farther reaches. Through the ages, many have written fictional accounts of space travel, especially voyages from the Earth to the Moon. The seventeenth century German astronomer Johannes Kepler, who calculated the orbits of the planets, wrote a fantasy of spaceflight and felt that his book could serve as a useful guide for people settling on the Moon. Space travel romances like Jules Verne's *From the Earth to the Moon*, published in 1856, described many kinds of spaceships. One of these craft consisted of a projectile fired from a gigantic cannon. Unfortunately, if a projectile were shot in this manner, the acceleration would kill the passengers. Even H.G. Wells, who was well known for his stories of space travel, such as *The First Men in the Moon* and *The War of the Worlds*, could not come up with a workable method for spacecraft propulsion.

Many ideas also were suggested for communicating with the inhabitants of Mars, including planting forests and grain fields in continent-sized geometric patterns that the Martians could easily identify as an indication of intelligent life. Today, fictionalized spaceflight is a booming component of the entertainment industry. For the most part, however, no attention is given to either physical principles or practicality.

The physical principles of spaceflight were laid down by the eighteenth century English physicist Sir Isaac Newton. A rocket travels, not by the push of the hot gases against the air, but in accordance to Newton's Third Law of Motion, which states that for every action there is an equal and opposite reaction. The expanding gases push equally against the combustion chamber walls and the outside, forcing the gases and the rocket to move in opposite directions, thus allowing the rocket to operate in the vacuum of space.

In order for the rocket to accelerate, Newton's Second Law of Motion must be invoked, which states that when a force acts on a moving body, the rate of change of momentum is proportional to the size and direction of the force. The greater the exhaust speed of the gases, or the *thrust*, the greater the speed of the rocket. Thrust is measured in pounds, and to lift a thousand-pound rocket requires more than 1,000 pounds of thrust.

Once the rocket is launched, Newton's First Law of Motion comes into play. This law states that every body remains at rest or in uniform motion in a straight line unless acted upon by an outside force. The force of gravity acting on a rocket will keep it from traveling in a straight line, causing it to curve back to Earth (FIG. 1-2). In order to achieve orbit, a rocket must reach such a velocity that centrifugal force (FIG. 1-3) overcomes the force of gravity. This is called the *escape velocity*, and for the Earth, it is about 25,000 miles per hour. Many rockets are launched from west to east to take advantage of the Earth's rotation, which gives an extra boost to the rocket (FIG. 1-4). Some satellites have elliptical orbits, taking them closer to the Earth at one point, then farther away at another (FIG. 1-5).

The next important step toward space travel was an understanding of the planetary motions

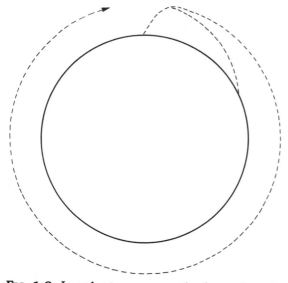

FIG. 1-2. In order to overcome the force of gravity and achieve orbit, a spacecraft must obtain an escape velocity of about 25,000 miles per hour.

around the Sun (FIG. 1-6). The planets do not orbit the Sun in perfect circles, as suggested by the Copernican system; instead, their orbits form ellipses with the Sun at one focus, as proposed by the German astronomer Johannes Kepler in

1609. The Kepler system allowed astronomers to accurately fix the orbits of the planets relative to each other, but it could not provide the actual distances of the planets from one another or from the Sun.

The relative distances of the planets from the Sun were mathematically expressed by the German mathematician Johann Titius in 1766 and became known as the Titius-Bode Law. In the series 3, 6, 12, 24, and so on, each number is twice its predecessor. Now, if 0 is placed at the beginning and 4 is added to each number, the series then becomes 4, 7, 10, 16, 28, 52, 95, 196 . . ., the Earth being number 10. With the exception of 28, which is about where the asteroid belt lies between Mars and Jupiter, the series fairly closely approximates the relative distances of the planets from the Sun.

In the late eighteenth century, an accurate measurement of the parallax of Venus as it passed between the Earth and the Sun established its absolute distance. Once the actual distance between two planets was known, the distances to all the rest of the planets could be readily calculated because their relative positions were already known.

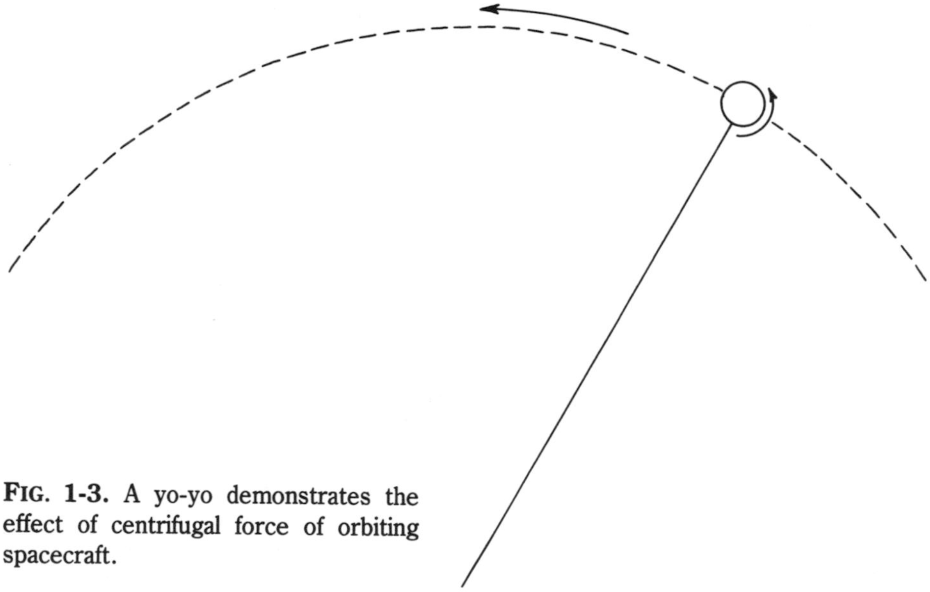

FIG. 1-3. A yo-yo demonstrates the effect of centrifugal force of orbiting spacecraft.

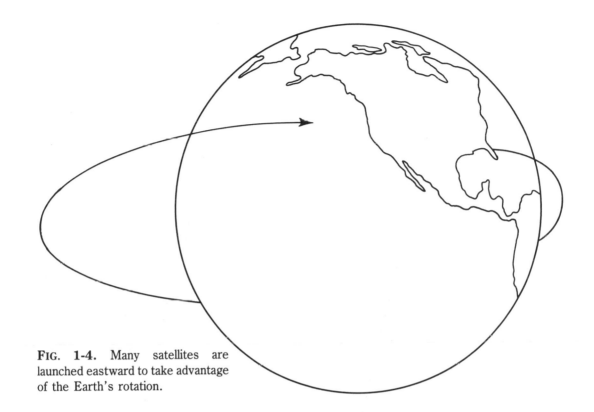

FIG. 1-4. Many satellites are launched eastward to take advantage of the Earth's rotation.

THE LIQUID-FUELED ROCKET

By the late nineteenth century, it was becoming more apparent that some form of rocket propulsion was needed for interplanetary travel because the rocket was the only known engine that could operate in the vacuum of space, where it would also be the most efficient with no air drag to slow it down. Solid-fueled rockets were considered unsuitable for space travel because of their explosive nature, and they were difficult to control, making them inaccurate and unreliable devices. In liquid-fueled rockets, however, fuel and oxidant are stored separately and fed into a combustion chamber at a controlled rate. The reaction produces hot gases that provide the propulsive force.

In 1883, the Russian teacher and scientist Konstantin Tsiolkovsky laid down the theoretical foundations for rocket propulsion. In 1903, he de-signed a rocket that used liquid hydrogen and liquid oxygen. Although never built, it was a highly workable design.

In 1908, the American physicist Robert Goddard built his first liquid-fueled rocket engine. In Auburn, Massachusetts, on March 16, 1926, he launched the world's first successful liquid-fueled rocket (FIG. 1-7), which burned kerosene and liquid oxygen and obtained a height of 184 feet. On July 17, 1929, Goddard launched a weather rocket, containing a barometer and a thermometer, as well as a camera to record their readings. The rocket rose 90 feet and traveled 171 feet before crashing to the ground, but the instruments landed safely by parachute. Goddard also wrote many scientific articles, including one on sending a rocket to the Moon. Although he was generally unknown in the United States, Goddard's works on rocketry were widely read in Europe.

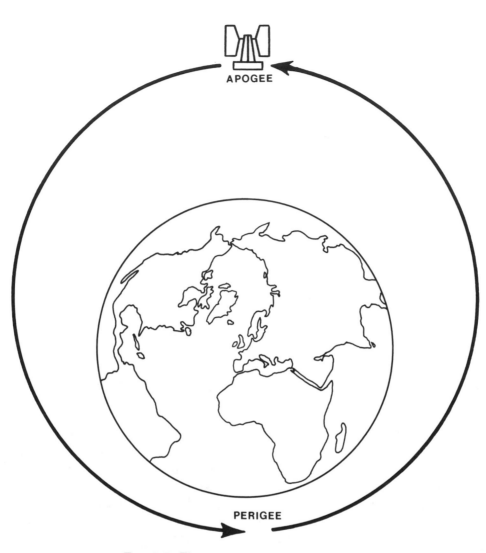

APOGEE

PERIGEE

FIG. 1-5. The apogee and perigee of a satellite.

The Germans were the most energetic about rocketry and space travel. By the late 1920s, several rocket clubs sprang up throughout that country to further the cause of space exploration. The German scientist Hermann Oberth developed formulas for calculating the energy requirements for a journey to the Moon and beyond, and his work is considered a cornerstone of astronautics. Another German, Walter Hohmann, worked out the various orbits rockets must obtain to reach the Moon and the other planets. During the depression of the 1930s, German rocket enthusiasts received funding from the military to continue their research and experimentation.

Although rockets have been used in warfare throughout the past millennium, including in such battles as the British siege of Fort McHenry in the War of 1812 (which inspired Francis Scott Key to write "The Star Spangled Banner"), it was the Germans who brought such weapons out of the archaic into the realm of modern weaponry. The German army began by organizing a research

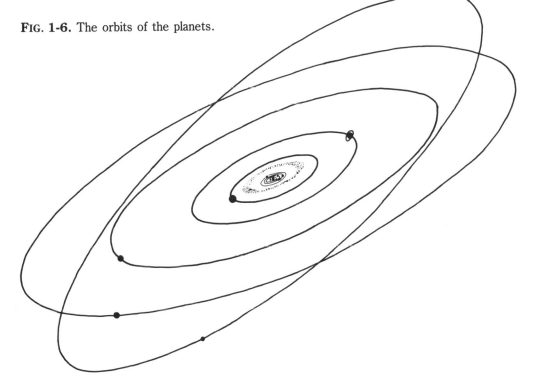

FIG. 1-6. The orbits of the planets.

facility at Kummersdorf, 60 miles south of Berlin. In December 1934, the army launched its first successful liquid-fueled rocket from Borkum Island on the Baltic Sea. The rocket was 4.6 feet long, 9 inches wide, and about 300 pounds in weight. It burned a mixture of ethyl alcohol and liquid oxygen, was the first rocket to use gyroscopic control for guidance, and obtained a record altitude of 6,500 feet.

Prior to the onset of World War II, a new rocket facility was built by the Germans at Peenemunde on the Baltic Sea and was headed by Wernher von Braun, who became director of the American space program after the war. The rocket became an integral part of Adolf Hitler's arsenal of wonder weapons, which he hoped would win the war for Germany.

A variety of rockets were designed and tested by the Germans, ultimately leading to the V-2 (FIG. 1-8), a highly advanced rocket that was 47 feet high and 5.5 feet wide, and weighed 14 tons fully loaded. It could carry a one-ton warhead of high explosives a distance of 200 miles, achieve an altitude of over 50 miles, and obtain a speed of Mach 5, or five times the speed of sound. The rocket used 4 tons of alcohol and 5 tons of liquid oxygen and developed a thrust of over 50,000 pounds. Its accuracy was very limited, however, and was mainly used as a terror weapon against the British toward the end of the war. A total of some 4,000 V-2s were launched, with about half used against London, England, alone.

In addition, the Germans had on the drawing boards a design for a three-stage rocket that could strike as far away as New York City. They also designed a pressurized capsule to be fitted on top of a V-2 rocket to carry the first man into space.

After the war, Germany was partitioned among the victors, and the United States and the Soviet Union searched the country for unlaunched

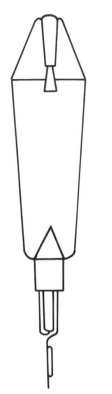

FIG. 1-7. Robert Goddard's first liquid-fueled rocket.

V-2 rockets and rocket scientists in their respective zones of occupation. Over one hundred scientists and technicians under Wernher von Braun were sent to work in the United States, while other scientists and technicians under Helmut Grottrup were captured and taken to the Soviet Union.

In the United States, some 60 V-2 rockets were launched from White Sands Proving Grounds in New Mexico. They carried a variety of payloads, including scientific instruments to probe the upper reaches of the atmosphere. A V-2 used as a booster stage for the smaller, solid-fueled Wac Corporal rocket set a world's altitude record of 244 miles on February 24, 1949. Cameras were carried in some of the rockets, and for the first time, it was possible to view photographs of the Earth taken from high altitude. Also, a live monkey was sent aloft to test the effects of acceleration

during liftoff. Unfortunately, after surviving the worst of the ordeal, the monkey was killed on impact when the parachute on its space capsule failed to deploy.

The Soviets rebuilt Peenemunde and the manufacturing plant at Nordhausen in East Germany and manufactured V-2 rockets there and also at a new facility outside Moscow. The Soviets built 30 serviceable V-2s and made significant improvements that effectively doubled the range of the rockets. The rockets were launched at the Kapustin Yar missile range, 75 miles southeast of Volograd (Stalingrad) and carried a variety of payloads, including warheads and scientific instruments. The Soviets also built a new category of much larger rockets. One rocket was 8 stories high and 9 feet wide, weighed 75 tons, and had a payload capacity of nearly 7000 pounds. The rockets were the forerunners of the Soviet long-range ballistic missiles that touched off an arms race with the United States. This fierce competition between the two superpowers produced spectacular results, culminating with the landing of a man on the Moon in less than 25 years after World War II.

THE SPACE RACE

Man's first serious venture into space began with rocket-powered aircraft, starting with the X-1, which set altitude and speed records in the late 1940s. These tests culminated with the X-15 (FIG. 1-9) which, prior to the space shuttle, was the closest thing to a winged spacecraft that was ever built. The X-15 was launched at high altitude from a bomber and was designed to operate over 50 miles above the Earth at speeds of over six times the speed of sound.

The X-15 contributed greatly to the manned space program by testing conditions that would be encountered by the first manned rockets. Because the air was so thin at these altitudes, normal aerodynamic controls were useless, and small rockets in the nose and wings were fired to control attitude and flight path. There were several proposals to launch an X-15 into orbit, using a carrier rocket, but the manned ballistic missile approach was more favorable and eventually led to Project Mercury. Ironically, in order to reduce the costs of manned space missions, the idea of a rocket-propelled aircraft was reawakened in the 1980s with the development of the space shuttle. Unfortunately, al-

most the entire American space effort was riding on the space shuttle program. When *Challenger* exploded on January 28, 1986, NASA was practically grounded until the problem with the seals on the solid-fueled booster engines was corrected.

The space race was born during the International Geophysical Year (IGY), July 1957 to December 1958, when scientists collected an

(Courtesy of U.S. Air Force)

FIG. **1-8.(a)** The launching of a V-2 rocket on May 10, 1946.

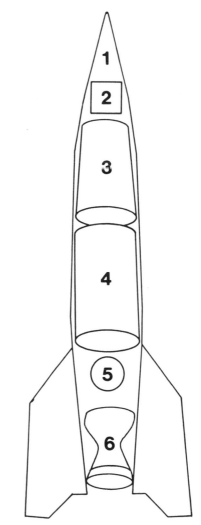

FIG. **1-8.(b)** A cross-section of a V-2 rocket: 1-warhead, 2-guidance system, 3-alcohol tank, 4-liquid oxygen tank, 5-turbine and pumps, 6-combustion chamber.

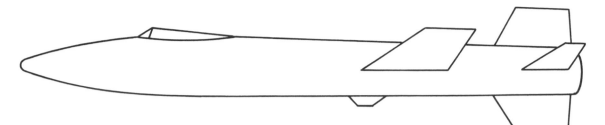

FIG. 1-9. The X-15 rocket plane.

intensive amount of data about the Earth. Both the United States and the Soviet Union pledged to launch scientific satellites as part of the IGY activities. The United States wanted its space program to be an entirely civilian affair to present to the world an image of peaceful intent and openness. Existing American military rockets, such as the Army's Redstone or the Navy's Viking (FIG. 1-10), would have made ideal boosters for launching satellites, but the government wanted an entirely new rocket for the project, called Vanguard (FIG. 1-11), with scientific responsibilities carried out by the National Science Foundation. In contrast, the space activities of the Soviet Union were controlled by the military and conducted in secrecy.

The American space effort was hampered from the beginning because it was separated from the military ballistic missile program and therefore deprived of many flight-proven components and much needed assistance. As a result, the Soviet Union, which was not hampered from making space exploration a military affair, was able to make several firsts in space.

On October 4, 1957, *Sputnik I*, the world's first artificial satellite (FIG. 1-12), was launched by a large Soviet ballistic missile into an eccentric orbit between 140 and 580 miles above the Earth. The satellite had a diameter of about 20 inches, weighed 184 pounds, and made a complete orbit in little over 100 minutes. It transmitted radio signals that carried information about the satellite's internal temperature. The signals also were used to determine its position. The transmitter worked

(Courtesy of U.S. Navy)

FIG. 1-10. Launching of the Viking 4 rocket on May 11, 1950.

FIG. 1-11. Vanguard leaves its launch pad at Cape Canaveral, Florida in October 1957.

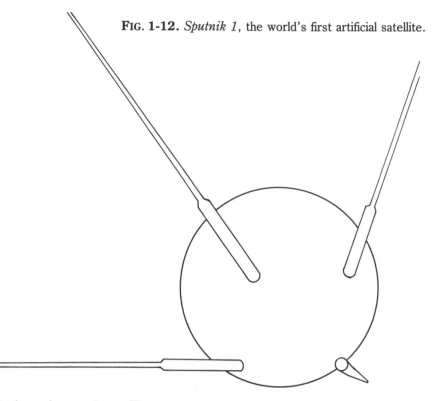

FIG. 1-12. *Sputnik 1*, the world's first artificial satellite.

for three weeks until the batteries ran down. The satellite remained aloft for three months until January 4, 1958, at which time its decaying orbit caused it to reenter the Earth's atmosphere and burn up.

Sputnik II, a considerably larger spacecraft weighing over half a ton, was launched a month after *Sputnik I*, on November 3, 1957, and for the first time, a live animal—a dog named Laika—was carried into orbit. The purpose of this and succeeding launches with live animals was to study the effects of space on laboratory animals before humans were sent aloft. Unfortunately, the means for bringing space capsules down without burning up in the atmosphere had not yet been developed. Therefore, Laika could not be retrieved and died long before the satellite spiraled earthward.

Following a series of failures with the Vanguard rocket, the United States finally launched its first satellite, *Explorer 1*, on January 31, 1958.

Its success was attributed to a crash program that adapted the Army's Jupiter C rocket into a four-stage carrier called the *Juno* (FIG. 1-13). *Explorer 1* was a small satellite, weighing only 31 pounds, but it carried more scientific instruments than *Sputnik I*. One of these instruments was a Geiger counter that could measure cosmic rays and was essential to the discovery of the *magnetosphere*, a region in space surrounding the Earth that is composed of magnetic fields that trap charged particles from the Sun and from outer space. The satellite's highly elliptical orbit—from 225 to 1585 miles—allowed it to sample a large region in space. This feature was paramount to the discovery of two large magnetic belts that circled the Earth above the equator, named the *Van Allen belts* in honor of their discoverer, the American physicist James Van Allen.

Explorer 1 was followed shortly by *Explorer 2*, which failed to achieve orbit (not an unusual

FIG. 1-13. A replica of the Jupiter C rocket with the *Explorer 1* satellite at the Smithsonian Institute. (Courtesy of U.S. Air Force)

FIG. 1-14. Launching of America's first man in space on the Mercury-Redstone 3 from the Cape Canaveral launch site on May 5, 1961. (Courtesy of NASA)

occurrence in those days). *Explorer 3* was launched on March 26, 1958, to continue the exploration of the Van Allen belts.

The first man into orbit was Soviet Air Force Major Yury Gagarin, who made one trip around the Earth in *Vostok 1* on April 12, 1961. The five-ton space capsule circled the Earth between 100 and 200 miles altitude and came down on dry land within the Soviet Union after being aloft for nearly 2 hours. Less than a month later, on May 5, 1961, America's first astronaut, Navy Commander Alan Shepard, made a suborbital flight on board *Mercury 3* (FIG. 1-14). The spacecraft reached an altitude of 116 miles and came down in the Atlantic Ocean, 300 miles east of Cape Canaveral, Florida, after being aloft for about 15 minutes.

Three months later, on August 6, 1961, the Soviets launched a second Vostok spacecraft, carrying Herman Titov seventeen times around the Earth in little over 25 hours.

Although the Soviet Union was unquestionably ahead of the United States in space, their triumphs spurred Americans on, and on May 25, 1961, President John F. Kennedy vowed to put a man on the Moon before the decade was out.

The United States did not make its first manned orbital flight until February 20, 1962, when Marine Lt. Col. John Glenn aboard *Mercury 6* stayed in orbit for nearly five hours. During this time, Glenn took pictures and gave a graphic narration of what he saw on Earth.

2

Probing the Planets

EVEN though man had taken the first steps into space, there still remained a whole universe to explore. Because of its nearness, the Moon was the next rung up the ladder of achievement. The Soviets were the first to send spacecraft to the Moon as a consequence of their development of large rocket boosters to launch heavy nuclear warheads, which had not been miniaturized to the same extent as those of the United States. On January 2, 1959, *Luna 1* flew within 5,000 miles of the Moon, and 2 months later, the United States flew a similar mission with *Pioneer 4*.

The first probe to crash-land on the Moon was the Soviet *Luna 2* on September 12, 1959, and two months later, *Luna 3* took the first pictures of the far side of the Moon. After a series of failures in 1962, the United States made dazzling successes in 1964–65 with Rangers 7, 8, and 9, which sent back thousands of pictures of the lunar surface. In February 1966, after four unsuccessful tries, the Soviets soft-landed a spacecraft on the Moon. The United States followed shortly with its Surveyor missions and the Lunar Orbiter series, which paved the way for the six Apollo spacecraft to carry men to the Moon and back. Although not considered an achievement that would further the welfare of mankind, man's journey to the Moon at least proved that modern technology could triumph over seemingly insurmountable problems (TABLE 2-1).

LUNAR LANDERS

The Soviet probe *Luna 2*, launched in September 1959 and the first man-made object to hit the Moon, sent back data that indicated there was no significant magnetic field on the Moon. Either the Moon does not have a liquid metallic core as does the Earth because it cooled and solidified, or it rotates too slowly to generate an appreciable magnetic field. The Moon completes a single rotation every 27.5 days, which also happens to be its orbital period around the Earth; thus the

TABLE 2-1. Summary of Major Space Probes.

PROBE	YEAR	MISSION
Mariner 2	1962	Fly-by of Venus; first probe to any planet
Mariner 4, 6, 7	1965 1969 1969	Fly-by missions to Mars
Apollo 8	1968	Astronauts circle the Moon and return to Earth
Apollo 9	1969	First astronaut landed on the Moon
Mariner 9	1971	Orbiter of Mars
Apollo 17	1972	Last of six Apollo missions to carry people to the Moon
Mariner 10	1974	Orbited the Sun, allowing it to pass Mercury several times; fly-by-mission to Venus
Pioneer 10, 11	1973 1974	First close-up views of Jupiter
Venera 8, 9, 10	1972 1975 1975	Soviet landers on Venus (operative about one year each due to harsh surface conditions)
Viking 1, 2	1976	Orbiters and landers of Mars
Voyager 1, 2	1979 1980	Fly-by of Jupiter Fly-by of Jupiter
Voyager 2	1986 1989	Fly-by of Uranus Expected fly-by of Neptune

Moon keeps the same side facing the Earth at all times.

After the success of *Luna 3*, which photographed the far side of the Moon as it approached within some 4,000 thousand miles, the Soviets then had a series of failures with Lunas 4 through 8, trying to make a soft landing on the Moon. They were finally successful on February 3, 1966, when *Luna 9* landed safely on the Moon and sent back close-up pictures of the lunar surface. Two months later, *Luna 10* was the first spacecraft to be placed into orbit around the Moon and was instrumented to study the Moon's magnetic field, micrometeoroid activity, and subatomic particles from the Sun.

America's first moon shot, *Ranger 1*, was launched from Cape Canaveral on August 23, 1961, but instead of going to the Moon, it went into a low orbit around the Earth and burned up in the atmosphere a week later. *Ranger 2*, launched three months later, suffered a similar fate. *Ranger 3*, launched on January 26, 1962, and programmed to hard-land on the surface of the Moon, missed the Moon by more than 22,000 miles. A computer on *Ranger 4* malfunctioned on its way to the Moon, and the probe crash-landed without sending back any useful data. On October 18, 1962, *Ranger 5* was well on its way to the Moon when it lost all its electrical power, and the dead probe sped 450 miles past the Moon and went into orbit around the Sun.

After a delay of over a year, *Ranger 6* was launched on January 30, 1964, and although the spacecraft was on course and would impact where it was supposed to, the six television cameras failed to turn on, and the probe crashed blindly into the Moon. *Ranger 7* was launched on July 28, 1964, took nearly 70 hours to reach the Moon, and crash-landed at a speed of almost 5,000 miles per hour. Before being destroyed on impact, however, the 800-pound spacecraft sent back more than 4,000 close-range pictures of the lunar surface. The cameras were first switched on when the probe was about 1,300 miles away, and for a quarter of an hour, detailed pictures of the Moon's surface were sent back until almost the very moment of impact.

Surveyor 1 (FIG. 2-1), launched on May 30, 1966, made a soft landing on the Moon's surface on June 2. Before power failed 1½ months later, it sent over 10,000 pictures back to Earth. It was followed on August 14 by the first of five Lunar Orbiters which sent back pictures of the lunar surface (FIG. 2-2) in search of potential landing sites for the Apollo missions. The Orbiters also provided much needed scientific data, including the discovery that the Moon's gravitational field was not uniform because of the existence of mass concentrations of dense material beneath the surface.

Surveyor 3 carried a remote-controlled mechanical digger to bring the lunar soil within full view of the cameras. *Surveyor 5* carried a portable chemistry laboratory to chemically analyze the soil. It determined that the Moon's surface layer was composed largely of basalt, similar to the gray volcanic rock found on Earth. Surveyors 6 and 7 continued work on the lunar surface and, together with the other Surveyors, proved that the surface of the moon was strong enough to support Apollo landing craft.

On July 20, 1969, Neil Armstrong and Edwin E. Aldrin, Jr. were the first to set foot on a planetary body other than the Earth (FIG. 2-3). This lunar space odyssey began on December 21, 1968, when *Apollo 8* became the first manned spacecraft to go to the Moon and return safely to Earth. The next manned mission, *Apollo 9*, launched on March 3, 1969, remained in Earth orbit to test out docking procedures with and test-firing of the lunar module. *Apollo 10*, launched on May 18, was sent into lunar orbit to test the lunar module's descent propulsion system and brought astronauts within 10 miles of the surface of the Moon. *Apollo 11* took off on July 16 and reached the Moon after coasting in space for four days. During the flight, the command and service modules separated from the final booster stage and turned around to dock with the lunar module (FIG. 2-4). At a distance of roughly 40,000 miles from the Moon, the spacecraft entered the *neutral zone*, beyond which the Moon's gravity took over. *Apollo 11* went into orbit about 70 miles above the Moon, and the lunar module, named *Eagle*, separated from the command-service module and descended to the surface. The astronauts collected rock and soil samples. After a stay of 21½ hours, the lunar module lifted off the Moon, linked up with the command-service module, and sped back to Earth.

VENTURE TO VENUS

Venus is called Earth's sister planet because its size and mass almost match that of the Earth. However, because of a veil of thick acid clouds, which totally obscure the surface from view, nothing was really known about this mysterious planet until probes were sent there beginning in 1961. Going to Venus was a greater challenge than traveling to the Moon because even at its closest approach, Venus is still a hundred times farther

(Courtesy of NASA)

FIG. 2-1. A full-scale mockup of a Surveyor spacecraft.

away. Actually, spacecraft are not sent directly to Venus as the orbits of Venus and Earth, bring planets close to each other, but rather, they are sent into intercepting orbits around the Sun and meet the planet at some prearranged destination. The same procedure is done for missions to Mars.

The first probe to visit another planet was the Soviet's *Venera 1*, launched on February 12, 1961. Unfortunately, after a few weeks into the flight, radio contact was lost, and although in all likelihood rendezvous was made, no data was sent back to Earth.

America's first Venus probe, *Mariner 1*, was launched on July 22, 1962, but the Atlas carrier rocket veered off course and had to be destroyed. *Mariner 2* was quickly made ready while the launch window was still open and was sent on its way August 27. It reached Venus on December 14 and flew within 22,000 miles of the planet. One of the most significant discoveries was that the average surface temperature on Venus is 860 degrees Fahrenheit, hot enough to melt some metals. There is no apparent magnetic field, and the planet takes 243 Earth-days to complete a

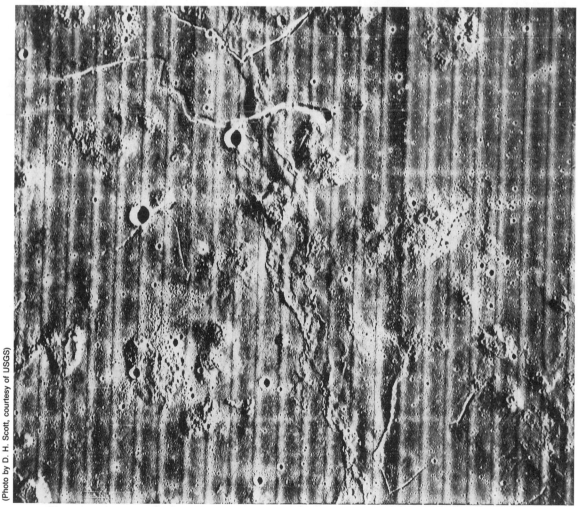

(Photo by D. H. Scott, courtesy of USGS)

FIG. 2-2. A Lunar Orbiter V photograph of the Marius Hills region of the Moon taken.

FIG. 2-3. Astronaut Edwin E. Aldrin, Jr. leaves the Lunar Module to explore the Moon's surface on July 16, 1969.

FIG. 2-4. The Stages of Apollo: 1-launch escape system, 2-command module, 3-service module, 4-lunar excursion module, 5-instrument unit, 6-S4B booster.

single rotation, which is *retrograde*, or backward with respect to the other planets.

Mariner 10, launched on November 3, 1973, passed close by Venus, taking pictures of the structure of the planet's upper atmosphere. Then by gravity assist from Venus, the probe sped on its way to Mercury (FIG. 2-5), which looked surprisingly similar to the Moon.

The first man-made object to touch another planet was the Soviet spacecraft *Venera 3*, which crash-landed on the surface of Venus on March 1, 1966. Information from Venera landers 4, 5, and 6, which parachuted through the thick atmosphere, showed that Venus contained 96 percent carbon dioxide with an atmospheric pressure upwards of 100 times that of the Earth— or equivalent to the pressure at 3,000 feet beneath the sea. The winds at the cloud tops blow from east to west at speeds over 200 miles per hour, and even 10-mile-per-hour winds on the surface have hurricane force because of the high pressure.

Two Soviet-French VEGA weather balloons, dropped in the middle of Venus' thick cloud layer in June 1985, found strong vertical winds and eddies the size of large weather systems on Earth. Under such harsh conditions of high temperatures and pressures along with hot sulfuric acid rains, Venus probes did not operate very long. In 1975, information about the chemistry of the crust came from Venera landers 8, 9, and 10, which measured the amount of radioactive elements in the soil and determined that geological processes are as much alive on Venus as on Earth. However, radar maps by Venus orbiters showed no evidence of a global network of spreading ridges and subduction zones, which are indicative of movable plates like those found on Earth and are responsible for most of the features of our planet's terrain.

The Venusian plain was photographed for the first time in March 1982 by the Soviet Venera 13 and 14 landing craft, which found a barren landscape strewn with flat and jagged rocks.

American Pioneer Venus spacecraft, which began orbiting the planet in December 1978, have sent four probes to the surface. One of the orbiters took high-resolution radar images of various landforms, including highlands with volcanolike features and large flat plains, which might have once been ocean basins before the planet became too hot and boiled away its water.

Venera 15 and 16, which have been in orbit since October 1983, have detected large circular features as much as 100 miles and more in diameter. These structures are relatively low in

(Courtesy of NASA)

FIG. 2-5. A simulated encounter with Mercury by the Mariner 10 spacecraft.

elevation, signifying that they might be *volcanic calderas*, or collapsed volcanic domes. One huge shield volcano in a region known as *Beta Regio* has a diameter of over 400 miles, considerably larger than the largest volcano on Earth. This is evidence that Venus is still volcanically active, which accounts for the great amount of sulfur dioxide in the atmosphere.

One of the most impressive features on Venus is an enormous canyon 900 miles long, 175 miles wide, and 8 miles deep. It is possibly the largest ditch in the Solar System, indicating that Venus might have had flowing rivers at one time.

THE MARS MISSIONS

Mars has always held a certain fascination for humans, ever since the first telescope was trained on the red planet. The famous early twentieth century American astronomer Percival Lowell even went so far as to suggest that the geometrical lines seen on the planet's surface were canals built to transport water from the polar ice caps to Martian desert cities on the equator.

If there were no other reason to go to Mars, it was important enough to investigate this strange phenomenon; for Mars was thought to be the best place outside the Earth to search for life in the Solar System. As with Venus, the Soviets took the lead, launching *Mars 1* on November 1, 1962. For several months while making rendezvous with Mars, the 2,000-pound probe functioned perfectly, yielding data about the solar wind and micrometeoroid activity. However, when the spacecraft was 65 million miles from Earth—the farthest a probe had ever been tracked at that time—it suddenly went dead, possibly because of a malfunction in the orientation system that keeps the large dish antenna pointed toward Earth.

America's first attempt to visit Mars—with *Mariner 3* on November 5, 1964—failed when a shroud protecting the spacecraft did not jettison, and the spacecraft lost power shortly after takeoff. *Mariner 4* was hastily made ready and launched on November 28. Two days later, the

Soviets sent up *Zond 2*, and it appeared that the two countries were neck to neck in a race to Mars.

After traveling some 325 million miles in eight months, *Mariner 4* passed within little over 6,000 miles of Mars on July 14, 1965, and sent back close-up pictures of the planet's highly cratered surface. The spacecraft found no magnetic field, and the atmospheric pressure was determined to be only about 1 percent of that of Earth. *Zond 2* probably passed close by Mars a few weeks later, but unfortunately, contact was lost when the spacecraft was still three months away.

The twin probes Mariner 6 and 7, launched on February 24 and March 27, 1969, respectively, came within 2,200 miles of Mars on July 31 and August 5, respectively. The planet was extensively photographed, and the atmosphere was found to be composed mostly of carbon dioxide. Temperatures as high as 70 degrees Fahrenheit were measured near the equator, while it was well over 100 degrees below zero at the poles.

Mariner 9 (FIG. 2-6), launched on May 30, 1970, went into orbit around Mars and photographed practically the entire surface between 1971 and 1972. When it first arrived, the probe encountered a planetwide duststorm that completely obstructed the view of the surface for several months. After the air finally cleared, the planet was found to be divided into two distinct hemispheres. The southern hemisphere was rough, heavily cratered, and traversed by huge channellike depressions. The northern hemisphere was smooth, only lightly cratered, and dotted with numerous extinct volcanoes (FIG. 2-7). The probe also revealed that the poles were covered with ice caps composed of frozen carbon dioxide, water ice, and windblown dust, and were as much as several miles thick.

The surface of Mars was photographed close up for the first time by the Viking Landers, which reached the planet in the early summer of 1976 and landed on opposite sides of the globe. The site of *Viking Lander 1* was a rolling plain that was littered with large boulders with several areas of exposed bedrock. Much of the landscape was covered by windblown sediment that indicated the winds blew from north to south. The site of *Viking Lander 2* (FIG. 2-8) was in the vast plains, which were remarkably flat and also strewn with boulders, but had no exposed bedrock.

The Viking Landers also were equipped to analyze soil samples for signs of life. Pictures from the Viking orbiters revealed that stream channels were abundant, indicating that Mars was once covered by oceans, or at least large lakes as much as 3 miles deep. A vast canyon, known as *Valles Mariners*, is 3,000 miles long, 100 miles wide, and up to 4 miles deep. It is evidence for a massive flash flood, possibly greater than anything experienced on Earth.

JOURNEY TO JUPITER

One of the greatest technological achievements of all times was the engineering of probes to be sent to Jupiter and beyond (FIG. 2-9). The United States began the exploration of Jupiter with

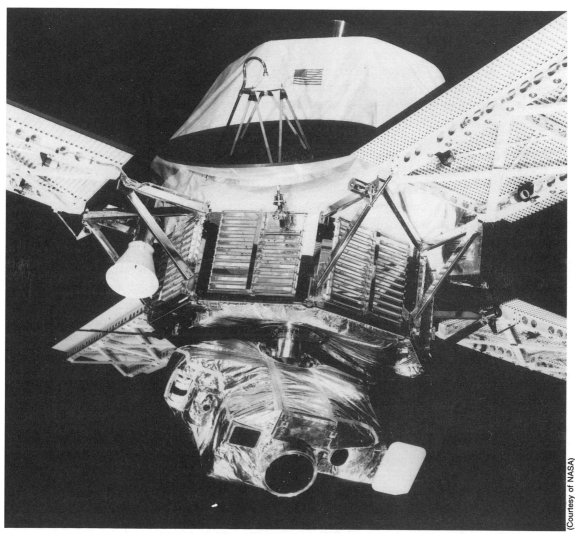

(Courtesy of NASA)

FIG. **2-6.** The 1971 Mariner Mars spacecraft.

the launching of *Pioneer 10* on March 2, 1972, and *Pioneer 11* on April 5, 1973. After traveling 600 million miles in eight months, *Pioneer 10* came within about 80,000 miles of Jupiter on December 3, 1973. It was followed by its twin probe a year and two days later.

On the way to Jupiter, the probes conducted experiments in interplanetary space, including mapping interplanetary magnetic fields, measuring the solar wind and galactic cosmic rays, and studying dust and micrometeoroids through the asteroid belt, which the probes had to cross to get to their destination. Actually, there was little danger of a collision with an asteroid, and the region was found to be relatively clear of all but the smallest particles. As the probes passed by Jupiter, they sent back hundreds of color photographs of the planet and its inner moons.

The pull of gravity from Jupiter accelerated *Pioneer 10* beyond the escape velocity of the Solar System, and in 1987, it became the first interstellar probe. Attached to the spacecraft was a plaque, designed by the American astronomer Carl Sagan, that describes us and our whereabouts in

(Courtesy of USGS)

FIG. 2-7. A mosaic of Olympus Mons, Mar's largest volcano, and surrounding countryside.

the Solar System, should an extraterrestrial stumble upon the probe during its long journey outbound.

The most remarkable feature of Jupiter is its multicolored bands of clouds, composed of condensed water droplets and ammonia, and the Great Red Spot, which is a gigantic storm system 14,000 miles across and 7,000 miles deep. The clouds are colored by substances brought up from below by convection and the interaction with ultraviolet light from the Sun which, incidentally, is only about 4 percent as bright on Jupiter as it is on Earth. The vigorous gyrations are produced by a strong Coriolis effect, caused by the high rotation rate of the huge planet, whose diameter is 11 times larger than Earth's, but whose day is only 10 hours long. The high rotation rate also produces strong winds with speeds upwards of 300 miles per hour, which are responsible for a great amount of turbulence in the cloud bands.

The atmosphere on Jupiter is composed mostly of hydrogen and helium in the same proportion found in the Sun. Had Jupiter grown somewhat larger, it could have ignited and Earth would be receiving light from two suns. The gases were swept into the condensing planet by its strong gravitational attraction, produced by a mass over 300 times larger than that of Earth. The planet has a solid, rocky core, about the size of Earth, that is covered by a massive ocean thousands of miles deep.

A considerable amount of information about Jupiter's moons was sent back by Voyager spacecraft that were launched in August and September

(Courtesy of NASA)

FIG. 2-8. A view of the Martian landscape from Viking 2.

1977. This was one of the most spectacular scientific achievements of all time. It began on March 5, 1979, when *Voyager 1* flew past Jupiter and returned close-up pictures of Jupiter's Galilean moons (FIG. 2-10),—the four inner moons named for Galileo, who discovered them in 1610. Perhaps the most dramatic discovery was the volcanic eruptions on Io, the innermost of the Galilean moons and the most volcanically active body in the Solar System. That moon is practically being pulled apart by a gravitational tug of war between Jupiter and its largest moon, Ganymede, causing Io to melt from the inside out. Io is pockmarked by numerous volcanic craters (FIG. 2-11) that spew a variety of sulfur compounds, which give the moon a brilliant red-orange appearance.

Europa is the brightest of the Jovian moons, with a reflectance ten times greater than Earth's moon because of the large amounts of water ice on the surface. Ganymede and Callisto, which are both about the same size as the planet Mercury, have a density of only about twice that of water and are composed in large part of water ice. The discovery of water ice on these moons indicates that there is a substantial amount of water in the Solar System, and that it was probably blown outward from the inner planets to the gaseous planets and beyond by the strong solar wind of the infant Sun.

SIGHTING SATURN

After its encounter with Jupiter, *Pioneer 11* was steered toward Saturn and rendezvoused with that planet in October 1979. Information from this probe indicates that the rings of Saturn are not continuous sheets as they appeared from Earth, but instead are made up of distinctive bands. At

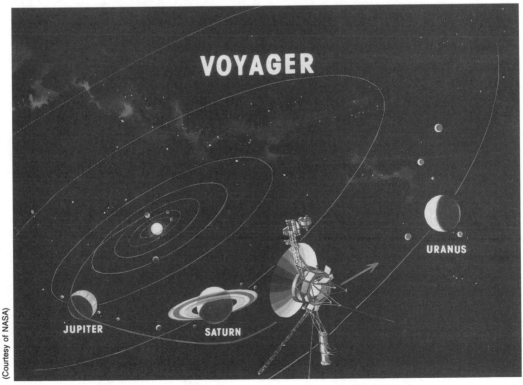

(Courtesy of NASA)

FIG. 2-9. Voyager spacecraft on its journey through the Solar System.

the same time, Voyagers 1 and 2, after receiving a gravity assist from Jupiter sped on their way to the ringed planet. *Voyager 1* caught up with Saturn in November 1980 and *Voyager 2* in August 1981, and both probes sent back breathtaking pictures of the planet, its rings, and its moons (FIG. 2-12).

Jupiter and Saturn are about the same size and composition, but Saturn's mass is only about one-third as much as Jupiter's, making it the only planet that could float in water. Saturn has similar colored bands of clouds and eddies because its rotational period is about the same as Jupiter's.

Saturn also has a rocky core that constitutes about 25 percent of the planet's mass, compared with only about 4 percent for Jupiter's core. Since Saturn is almost 1 billion miles from the Sun, most of its heat is generated internally by gravitational self-compression. This outward flow of heat is responsible for strong convection currents, which provide Saturn with many of its atmospheric features.

The Voyager spacecraft revealed the most bizarre set of moons found anywhere in the Solar System. Before their arrival, the known moons

(Courtesy of NASA)

FIG. 2-10. A collage of Jupiter and its Galilean moons Io, Europa, Ganymede, and Callisto photographed by Voyager 1.

of Saturn were seen as mere faint, tiny dots even through the most powerful telescopes. All but 1 of the 17 moons rotate synchronously, keeping the same side facing the planet, like the Earth's moon. All but two of the moons have orbits that are nearly circular and lie in the equatorial plane of the planet. The outermost moon, Phoebe, is the only known moon that orbits its mother planet in the opposite direction of all the other moons.

Every moon of Saturn has a mass less than twice the density of water, indicating that they are composed of rock and ice. Their high surface reflectance, between 60 and 90 percent, suggests that the moons are coated with ice. Iapetus, the second outermost moon, is half black and half white, and performs a disappearing act as it revolves around Saturn. Rhea and Dione are intensely cratered (FIG. 2-13), similar in appearance to the earth's moon. Tethys has a huge crater, two-fifths the diameter of the moon itself, and a branching canyon that is 600 miles long, 60 miles wide, and several miles deep.

Saturn's largest moon, Titan, was visited close up by *Voyager 1*, which was taken out of trajectory in order to obtain a better view at this mysterious moon. It was well worth the sacrifice for Titan was found to be the only known moon with a substantial atmosphere (FIG. 2-14), one that was even denser than Earth's. Moreover, the atmosphere resembles that of the early Earth

(Courtesy of NASA)

FIG. 2-11. A view of Io from Voyager 1, showing numerous volcanic craters.

and therefore Titan would seem to be the likeliest place in the Solar System to search for the precursors of life.

URANUS UNCOVERED

Following its close approach to Titan, *Voyager 1* headed for the outer fringes of the Solar System, unable to be redirected by Saturn's gravity toward Uranus. That left only *Voyager 2* to carry on with the exploration of the outer planets. The decision to continue on to Uranus (FIG. 2-15), which was not on the original itin-

erary, resulted in extensive in-flight engineering during the spacecraft's travel through inter-planetary space. Voyager's computers had to be reprogrammed to overcome malfunctions in the probe's radio receiver and camera tracking platform.

At this great distance from the Sun, over twice as far as Saturn (1.8 billion miles), Uranus and its moons are poorly illuminated. In addition, Uranus orbits the Sun on its side, requiring the probe to shoot through the orbital plane of its moons, instead of moving within the orbital plane

(Courtesy of NASA)

FIG. 2-12. A collage of Saturn and its major moons, with Dione dominating photographed by Voyager 1.

of the moons as with the other planets, thus providing less time for tracking the moons of Uranus. Also, coded instructions transmitted to the spacecraft and data sent back to Earth required over five hours round trip at this distance. As if these were not enough problems, the spacecraft's electrical supply, which is generated by the nuclear decay of radioisotopes, had declined to about 80 percent of capacity; therefore, some systems had to be shut down in order for others to operate. Regardless of these great obstacles, *Voyager 2* was able to return striking images of the pale blue planet and its 15 moons.

Some of the moons of Uranus exhibit the most extraordinary collection of terrain features ever seen in the Solar System. Miranda (FIG. 2-16) is

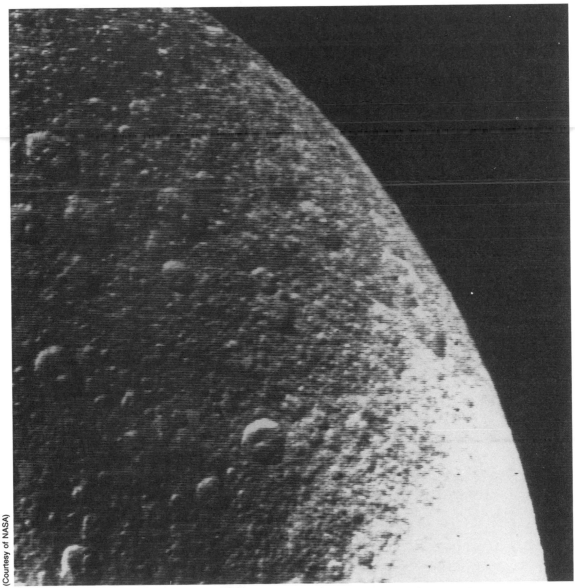

(Courtesy of NASA)

FIG. 2-13. A view of meteor impact craters on Rhea from Voyager 1.

the smallest and most exotic of the five major moons, and looks as though it were smashed by a colliding comet and reassembled with all its parts in the wrong place. The surface of Titania, the largest of the Uranian moons, bears dramatic evidence of global tectonics, including a complex set of rift valleys bounded by faults like cracks on a boiled egg. Some areas have only a few craters and appear to have been resurfaced by volcanic activity—not with hot lava, but a mixture of rock and water ice. Oberon is only slightly smaller than Titania, but shows only a few signs of geologic activity. Instead, it has many large meteor craters from a massive bombardment around 4 billion years ago. Because only the southern hemispheres of the moons were illuminated by the Sun,

much is still to be learned about this strange new world.

Voyager 2 continued toward Neptune, thought to be the twin of Uranus. If all goes well, in August 1989, the probe should send back images of Neptune and its moons.

FUTURE FRONTIERS

Perhaps the greatest challenge of America's space effort is the construction of a space station (FIG. 2-17), which should begin sometime after the space shuttle flies again, and possibly be completed by the mid-1990s. The $30 billion project will be the largest structure ever put into space and will be the home for up to eight astronauts, with living quarters, laboratories, and work areas.

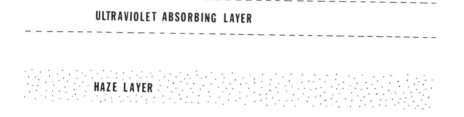

ULTRAVIOLET ABSORBING LAYER

HAZE LAYER

AEROSOL LAYER

FIG. 2-14. The atmosphere on Titan.

METHANE   CLOUDS

There are also plans for a permanent base on the Moon for astronautical research, and the space station will be an ideal rendezvous point for future manned missions to the Moon and Mars.

A one-way journey to Mars with our present rocket technology is expected to take up to six months. After making a soft landing on Mars, astronauts will conduct a complete scientific survey of the planet and return to Earth with rock samples.

In addition to the Mars mission, new unmanned probes like the Venus radar mapper and the Galileo mission to Jupiter will further extend our knowledge of our planetary backyard.

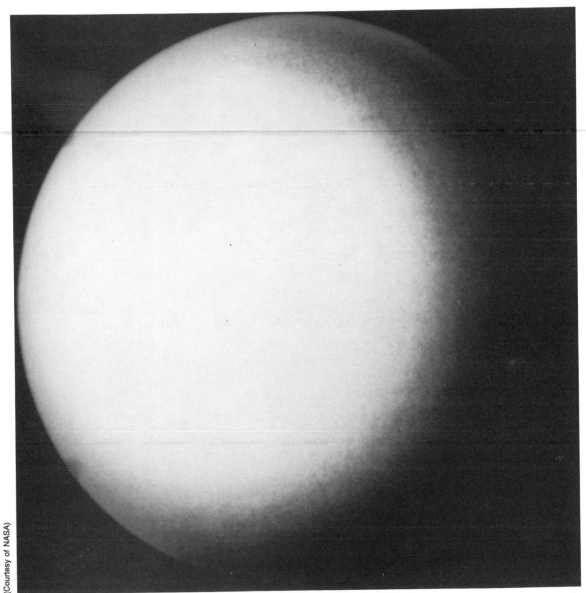

(Courtesy of NASA)

FIG. 2-15. A view of Uranus from Voyager 2.

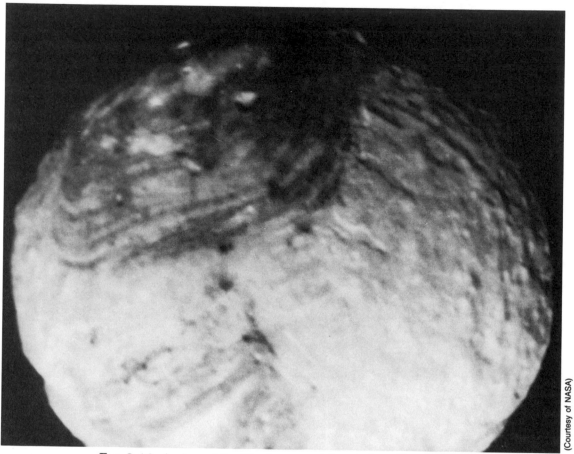

(Courtesy of NASA)

FIG. **2-16.** A view of the Uranian moon Miranda from Voyager 2.

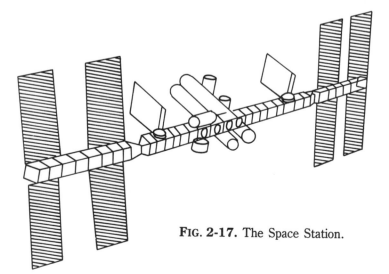

FIG. **2-17.** The Space Station.

3

Eyes in the Skies

AFTER meeting the challenges of exploring the Solar System, it was time to turn our attention to Earth, and apply the technology we gained from our adventures in space toward collecting data on weather and climate, on earth resources, and on air and water pollution for the benefit of all inhabitants of the planet. The first surveillance satellites were developed for the military, and were used to locate missile-launch sites and other military installations. Clouds were generally a nuisance to such operations because they obstructed the satellite's view of the ground. However, the ability to observe clouds from space eventually led to the development of weather satellites. Out of this technology grew the development of earth resource satellites, whose ability to resolve objects on the ground, called its *resolution*, was sufficient for most land-use applications.

The commercialization of space has led to an increase in the use of privately owned satellites to photograph Earth. The resolution of some of the new European satellites is approaching that of military surveillance satellites, which has military planners concerned over problems of national security. However, the refinement of these satellites is of vital importance for obtaining the best possible data about the Earth from space.

SPY SATELLITES

In August 1960, just three years after Sputnik sparked off the space race with the Soviet Union, the United States launched the first successful suborbital spy satellite, called *Discovery*. It made three passes over the Soviet Union, taking pictures along the way, and was recovered in the Pacific Ocean after its decaying orbit brought it back to earth.

Film capsules were later jettisoned from permanent orbiting satellites and recovered in midair by net-bearing aircraft. On January 31, 1961, a new spy satellite, called SAMOS for Satellite and Missile Observation System (FIG. 3-1), was launched successfully. Rather than use the precarious method of dropping film capsules out of the sky, SAMOS transmitted back images of photographs automatically developed on board the satellite.

Three infrared heat-seeking satellites, called MIDAS for Missile Detection Alarm System, were placed an equal distance around the world in geosynchronous orbits 22,300 miles above the equator. At this altitude, the satellites orbited Earth the same rate Earth rotates, or once every 24 hours. The satellites covered the entire landmass and the oceans of the world, and could detect any missile launched by sensing the infrared

heat produced by their rocket motors. This information was processed by computers on the ground and collated with other data to determine whether the launching was for peaceful purposes or was an attack on the United States.

Among the over 5,000 objects currently in orbit around Earth is an assortment of spy satellites, which are divided into two main categories: photoreconnaissance and ferrets. *Photoreconnaissance satellites*, such as the American KH-8, gave the best close-up views, but could only stay aloft for a month or so. The KH-9, also called Big Bird, could stay up for almost a year, and KH-11, which could operate for two years, transmitted high-resolution pictures in real time,

(Courtesy of U.S. Air Force)

FIG. **3-1.** A Samos space vehicle is mated with payload.

or live. The Soviet Union's photoreconnaissance satellites tend to be short-lived, single-mission devices. What the Soviets lack in technology they make up for in quantity, through a nonstop launch schedule. Of the 100 or so Soviet payloads sent into space each year, at least one-third are for military purposes.

Both superpowers also have satellites called *ferrets* that can monitor *signal intelligence* (SIGINT), which is radar activity and missile telemetry, and *electronic intelligence* (ELINT), which is radio communications. Ferrets can be placed in geosynchronous orbits, where they sit indefinitely over one spot, or they can be placed in lower orbits and gather information as they pass overhead.

COMMUNICATIONS SATELLITES

The world's first communication satellite also happens to be the Earth's own natural satellite: the Moon. Soon after World War II, on January 11, 1946, the U.S. Army Signal Corps bounced radar signals off the Moon in order to accurately determine its distance from the Earth. On November 29, 1959, Bell Telephone engineers relayed radio transmissions from Holmdel, New Jersey, to Goldstone, California, via this same natural satellite. During the 1950s, the U.S. Army also used the Moon as a reflector of radio waves when atmospheric disturbances caused existing channels of communications between the United States mainland and Hawaii to fail.

Because of this ability to bounce radio waves back to Earth, the first artificial communications satellite, appropriately named *Echo 1*, was nothing more than a 100-foot-diameter, metallic-coated balloon, launched 1,000 miles in orbit on August 12, 1960. The satellite provided a bonus to *geodesists* (scientists who measure the Earth). They used it to plot geographical locations with far greater accuracy than was possible using earthbound geodetic methods.

The first active communications satellite— one that received, amplified, and retransmitted radio signals—was a 15-pound battery-operated test pod called *Score*, attached to an Atlas booster rocket (FIG. 3-2) and launched on December 18, 1958. The satellite contained two radio receivers, a tape recorder, and two transmitters. It received a message from one location, stored it on tape, and when given the command signal, retransmitted it to another location.

(Courtesy of U.S. Air Force)

FIG. 3-2. An Atlas booster is used to launch a variety of payloads.

A second delay-repeater satellite, called *Courier*, was launched on October 4, 1960, only this time the 500-pound satellite was detached from its carrier vehicle and was powered by arrays of solar cells on its spherical surface.

The first commercial communications satellite, called *Telstar*, was built by the American Telephone and Telegraph Company (AT&T) and launched on July 10, 1962. It was designed to test the feasibility of relaying live television broadcasts across the Atlantic Ocean to Europe. The disadvantage of these early communications satellites was that their low orbits required users to wait until the satellite was overhead before they could send their broadcasts.

It was not until the development of large carriers that could boost heavy payloads into *geosynchronous orbit*—the satellites remained constantly over one spot, above the equator—that communications satellites became fully functional around the clock. The first successful communications satellite to be placed into such an orbit was *Syncom* (FIG. 3-3), developed by NASA and the U.S. Department of Defense, and launched on July 23, 1963. Its success paved the way for international cooperation in space, with the organization of the International Telecommunications Satellite Consortium (Intelsat), and resulted in the launching of the Early Bird communications satellite on April 6, 1965. This satellite operated for about 3½ years and offered two-way voice communications across the Atlantic Ocean equal in telephone circuit capacity to the two existing transatlantic submarine cables. By 1969, full global coverage was achieved with six Intelsat communications satellites, each of which could carry upwards of 1,200 telephone calls at one time.

Some 100 nations presently are linked by over 12 such satellites, and the number of international telephone circuits made through satellites has grown considerably. The timely exchange of televised news and cultural events via satellite is now commonplace among most countries, and the world has shrunken now that people are better in touch with each other.

WEATHER SATELLITES

The first weather satellite, called *TIROS 1* for Television Infra-Red Observation Satellite, was launched from Cape Canaveral on April 1, 1960. The 270-pound satellite operated for only 78 days, but during that time, it produced over 20,000 pictures of Earth. This was an experimental craft, the first in a series of polar-orbiting weather satellites of the National Operational Meteorological Satellite System, and able to view the global atmosphere on a regular basis both day and night. The Earth rotates beneath the satellites as they circle in near polar, Sun-synchronous orbits approximately 500 miles altitude (FIG. 3-4).

The latest versions, designated *TIROS N*, are operated by the National Oceanic and Atmospheric Administration, and these satellites are appropriately named NOAA. The satellites make about 16,000 global atmospheric soundings daily to the National Weather Service. They provide valuable information about weather conditions over ocean areas where conventional data are lacking. The newest NOAA satellites provide capabilities for search and rescue missions, ozone mapping, and meteorological data for the World Weather Watch Program.

The Nimbus weather satellite (FIG. 3-5) was launched into polar orbit on August 28, 1964, and carried advanced sensors that provided the first nighttime pictures of cloud cover. After only a month of operation, its two large solar paddles would no longer function, and the satellite lost power.

A second Nimbus satellite was sent into orbit on May 15, 1966. In addition to the same sensors as its predecessor, it carried equipment to measure solar radiation reflected by the Earth into space. Later versions of Nimbus weather satellites carried a variety of experimental packages and operated as testbeds for the Earth Resource Observational Satellite (EROS) system.

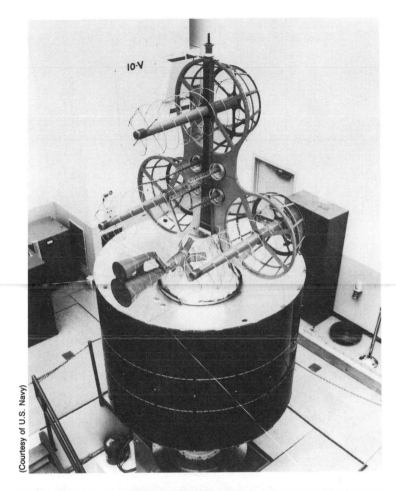

(Courtesy of U.S. Navy)

FIG. 3-3.(a) A synchronous communication (syncom) satellite called Marisat provides telecommunications services for the U.S. Navy and the commercial maritime industry.

FIG. 3-3.(b) A map showing areas under surveillance of Marisat.

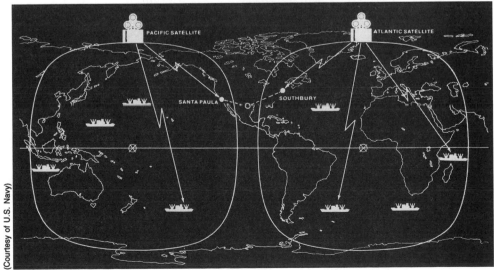

(Courtesy of U.S. Navy)

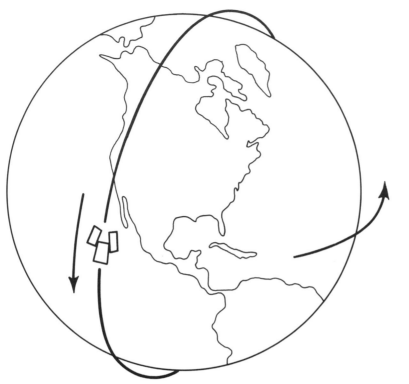

FIG. 3-4. The Earth rotates beneath a polar-orbiting satellite.

A continuous surveillance of weather in the United States required a different type of satellite: one with an orbital speed exactly matching the rotation of Earth so that it hovered constantly over the same point on the Earth's surface. This was accomplished by sending the satellite into a geostationary orbit about 22,300 miles over the equator. The first experimental geostationary weather satellites were launched in 1966 and 1967, providing the technological base for the Geostationary Operational Environmental Satellites (GOES), operated by NOAA. The first GOES prototype was launched on May 17, 1974, and a second on February 6, 1975. GOES 5, positioned to cover the East Coast of the United States, and GOES 6, centered over the central Pacific, were launched in the early 1980s. Each satellite viewed almost one-third of the Earth's surface and provided images every 30 minutes, day or night. In 1984, after three years of service, GOES 5's weather tracking camera failed. As a result, GOES

6 had to do double duty, so it was moved to a more central position to cover the entire United States by itself. In May 1986, GOES 7 was launched to replace GOES 5, but the Delta rocket used as a carrier malfunctioned shortly after liftoff and had to be destroyed along with the satellite. The eighth and final satellite in the GOES series was successfully launched on February 25, 1987 (FIG. 3-6).

The satellites are particularly important for early warning of tropical storms. Almost every storm system that strikes the United States forms in the North Pacific off the southern coast of Alaska or in the North Atlantic off the west coast of Africa (FIG. 3-7). Weather satellites also provide nearly continuous coverage of the atmospheric environment that breeds tornadoes, squall lines, and other local severe convective storms. Telltale signs of strong convection often can be detected in satellite imagery before severe storms develop. After they do develop, the

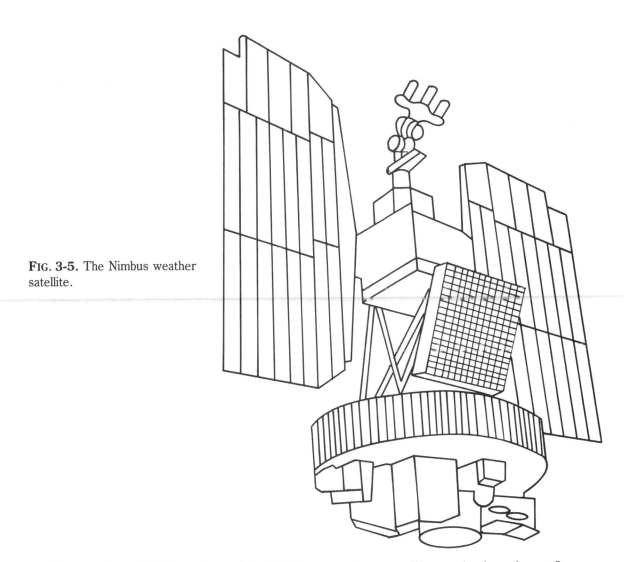

FIG. 3-5. The Nimbus weather satellite.

satellites monitor their life cycles and track their movements. When necessary, the National Weather Service gives timely warnings of impending danger.

The satellite imagery also is used to estimate rainfall amounts for flash flood warnings. In the winter, the satellites monitor mountain snowpacks and river ice jams to warn of possible flooding. Satellite images are used to monitor the southward progress of freezing surface temperatures and provide warnings for citrus growers. The satellite sensors are sensitive to high heat sources and have been effective in pinpointing wild fires in remote areas. The smoke from forest fires, as well as the ash cloud from volcanoes (FIG. 3-8) is also easily detected. The satellites monitor sea surface temperature, which is not only of value to fishermen but can provide early warning of unusually strong El Niño currents, which have a major impact on the weather the world over.

EARTH RESOURCE SATELLITES

Out of the basic design for the Nimbus weather satellites came a series of earth resources satellites. The first, called Earth Resources Technology Satellite, or ERTS 1, was launched

FIG. 3-6. The launch of GOES 8 with a Delta rocket on February 25, 1987.

on July 23, 1972. It was later renamed Landsat 1 and joined by Landsat 2 in January 1975. Landsat 1 failed after 5½ years of service, and a third Landsat was launched in March 1978 to take its place. The satellites employed a system of three television-type cameras and a multispectral scanner, which uses an oscillating mirror scanning system to image the ground. The multispectral scanner images the Earth in green and red visible wavelengths and two invisible near-infrared wavelengths. The blue wavelength is not used because it is scattered and absorbed by the atmosphere, giving objects a hazy appearance.

The next generation of satellites began with Landsat 4 (FIG. 3-9), launched in July 1982. After only one year of operation, however, its performance began to deteriorate from loss of power. The latest version, Landsat 5, was launched on March 1, 1984, to take its place, and as of November 1988 was the only fully functional satellite in the series.

In addition to a multispectral scanner, Landsats 4 and 5 were equipped with a Thermatic Mapper, which has a higher resolution and therefore can image in greater detail. The multispectral scanners can resolve features on the

FIG. 3-7. A Gemini 5 photograph of a storm off the coast of northwest Africa.

(Courtesy of NASA)

ground as small as 270 feet, or about the size of a football field, and the imagery covers a swath 115 miles wide. The Thermatic Mapper has a 98-foot resolution and can view the Earth in 7 spectral bands in the visible and infrared regions.

These Landsat satellites operate in a polar orbit at an altitude of about 435 miles, and circle the planet about every 100 minutes. This type of orbit was chosen so that Landsat can keep pace with the Sun as the Earth spins beneath the satellite, giving it a Sun-synchronous orbit. As a re-

sult, Landsat 5 always passes over a certain point on the ground at the same local time—between 9:30 and 10:00 A.M.—and returns to the same point every 16 days. The advantage of this particular orbit is that the angle of the Sun is always the same, and differences in imagery taken on different days are more easily determined.

In September 1985, NOAA transferred operation of the Landsat program to the Earth Observation Satellite Company (EOSAT), a joint venture between RCA and Hughes Aircraft that

VOLCANIC ASH CLOUD ▶

(Courtesy of U.S. Navy)

FIG. 3-8. The ash cloud from the 1980 eruption of Mount St. Helens.

now runs Landsats 4 and 5 and plans to build and operate two similar satellites. The transfer of Landsat to the private sector resulted from an economic condition that arose in 1981 when NASA decided it could no longer afford to keep up with the research and development of new sensors for NOAA's satellite system. After NASA's withdrawal, NOAA was left with more responsibility than it could handle alone, and the federal government decided that a private corporation could market Landsat data more successfully.

Unfortunately, even with EOSAT operating the satellites, the problems are not over. Budgetary restraints called for only one new satellite, instead of the original two. In addition,

Landsat 5 which is almost at the end of its useful life, (which is generally five years) will not be replaced by Landsat 6 until late 1989. This could leave a gap in the data flow, and customers might need to turn elsewhere to obtain satellite imagery.

On February 21, 1986, the French successfully launched the remote-sensing satellite SPOT 1 (FIG. 3-10) aboard the European Space Agency's Ariane rocket. The launch dramatized the emergence of space remote sensing as a viable commercial enterprise. Images of Earth are marketed on a worldwide bases to customers from fishermen to oil companies. SPOT provides high-resolution panchromatic and multispectral images. The images cover four spectral bands in the visible

FIG. 3-9. An artist's rendition of Landsat 4, the first in a new series of earth-resource satellites using the Thermatic Mapper.

(Courtesy of NASA)

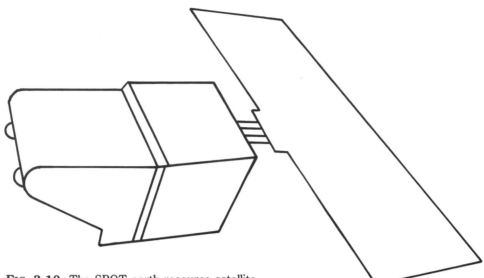

FIG. 3-10. The SPOT earth-resource satellite.

and near-infrared wavelengths and offer a spatial resolution of 33 feet, considerably better than Landsat 5. The improved resolution provides greater detail on the ground, giving customers the best commercially available views of Earth from satellite imagery.

ASTRONOMICAL SATELLITES

Ever since Galileo invented the telescope in the early seventeenth century, astronomers have been plagued with a turbulent atmosphere that causes their images to blur. Some wavelengths, like the far infrared, are impossible to view from the ground because of atmospheric interference; therefore, infrared astronomy must operate outside Earth's atmosphere.

On January 25, 1983, the Infrared Astronomy Satellite (IRAS) was successfully launched into orbit from Vandenberg Air Force Base, California. Its purpose was to survey the sky at infrared wavelengths that contain information on some of the most important emissions from planets, asteroids, comets, interstellar molecules, and newborn stars and galaxies. In addition, infrared astronomy is able to penetrate the interstellar clouds that obscure much of the Milky Way gal-

axy from optical astronomy. The useful lifetime of the satellite was limited to the amount of liquid helium it contained, which was used as a coolant for the detector and was slowly bled off to keep the telescope's temperature down so its own thermal emissions would not interfere with the object it was observing.

On November 22, 1983, after ten months of operation, the coolant ran out, leaving IRAS permanently blinded. During its useful life, however, the telescope recorded images of 98 percent of the celestial sphere, demonstrated the power of infrared astronomy from space, and set the stage for future IRAS missions. Although its sensors were out, IRAS's computers were still fully functional and were used as a testbed for reprogramming future astronomical satellites.

X-rays are another region of the electromagnetic spectrum that are difficult to observe from the Earth's surface because of absorption by the atmosphere. On November 13, 1978, the Einstein X-Ray Observatory (FIG. 3-11), an X-ray telescope, was launched into orbit. For the first time astronomers were provided with a telescope whose sensitivity to X-rays matched the sensitivities of optical and radio telescopes. It

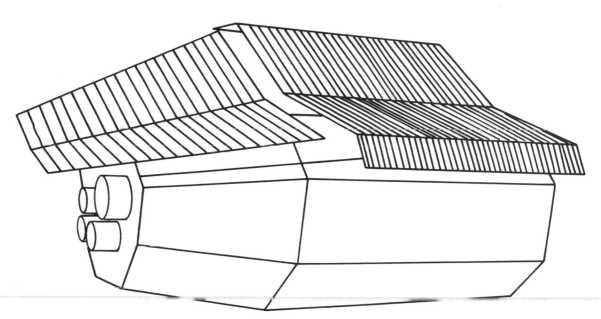

FIG. 3-11. The Einstein X-ray observatory satellite.

recorded at X-ray wavelengths every major class of astronomical objects that have been detected by the largest ground-based telescopes.

In its brief, but profitable, existence, information from the Einstein Observatory showed that normal stars emit far more X-rays than had previously been thought. It has greatly increased the number of known *binary star systems*, in which a large star has a highly dense companion such as a neutron star or a black hole. X-ray emissions provided a way to classify star clusters that was not possible with observations at other wavelengths. The telescope also recorded X-ray emissions from the most distant known *quasars*, which are starlike objects that are receding at nearly the speed of light and account for most of the X-ray background radiation. Before atmospheric drag took the satellite out of orbit in 1981, the Einstein Observatory clearly demonstrated the immense value of conducting astronomical investigations in the X-ray region. In the future, more sensitive telescopes will continue to make dramatic new discoveries.

Observation of the Sun outside the Earth's atmosphere was conducted by the Solar Maximum

Mission (Solar Max) satellite beginning in February 1980. Radiometers on board the satellite measured a decrease of *solar irradiance*, or brightness, of nearly 0.1 percent from 1981 to 1984. That is quite a drop for something that for centuries astronomers thought was so stable they called it the *solar constant*. Even such a small decrease in luminosity as this over a long period could have an effect on global climate.

The apparent decrease in solar output might be caused by a decrease in the number of sunspots since the beginning of the decade. The sunspots themselves do not block out solar energy streaming toward Earth, but instead are an indication of increased solar activity (FIG. 3-12). Sunspots appear dark because they obstruct the convective heat flow toward the surface of the Sun, while they uncover the hotter depths below and thus increase their radiation into space. The sunspot numbers peak about every 11 years and change their magnetic polarity about every 22 years. The 11-year cycle reached a minimum in 1975, a maximum in 1981, and a minimum again in the autumn of 1985. With the beginning of each new cycle, the Sun's irradiance increases. One interesting solar phenomenon that could be responsible for this increase in irradiance is an apparent expansion of the Sun, providing a link between the Sun's size and its solar output.

ORBITING LABORATORIES

The orbiting laboratory Skylab (FIG. 3-13), constructed out of leftover Apollo hardware, was sent into low orbit around the Earth on May 14, 1974. Skylab had an overall length of about 120 feet and a mass of about 100 tons. At the time, it was the largest man-made object in space. During its operation, over 500 man-days in space were devoted to solar astronomy, Earth observations, astrophysics, life sciences, and a variety of experiments in space, including manufacturing substances in zero gravity. The last crew

remained for 84 days, an American endurance record, testing the long-term effects of living without gravity.

The Skylab experience was the first time crews erected structures in space to repair the spacecraft and to test construction techniques for building a permanent space station. While waiting for the space shuttle, which was over a year behind schedule, to push it into higher orbit, intense solar activity increased atmospheric drag on Skylab, and it came down like a huge blazing meteor over Australia in July 1979.

Spacelab 1 (FIG. 3-14) was launched aboard the space shuttle on November 28, 1983, and was brought back down on December 8. The orbiting laboratory was built by the European Space Agency, and astronauts conducted scientific

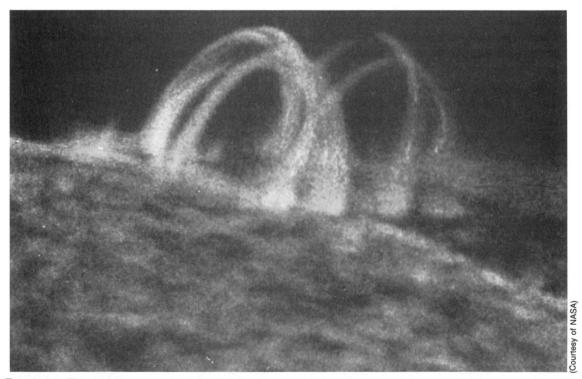

(Courtesy of NASA)

FIG. 3-12. The turbulent nature of the Sun is demonstrated by these magnetic loops observed by the Solar Telescope on Skylab.

experiments in the zero gravity of space. There were five major areas of scientific research: astronomy and solar physics, space plasma physics, atmospheric physics and Earth observations, materials science, and life sciences.

The laboratory contained 38 different experimental facilities that conducted more than 70 investigations. Some investigations produced immediate results, such as those in the life sciences, while other studies involved the collection of enormous amounts of *data* with enough information to fill over 100,000 encyclopedia volumes, requiring up to a decade to analyze. The greatest potential was shown in materials processing in space. Crystals grown on Earth must contend with the force of gravity, which dictates their shapes and sizes. Crystals grown in space, such as protein crystals, which are important for pharmaceuticals, can obtain a volume 1,000 times greater than those grown on Earth.

KILLER SATELLITES

The United States and the Soviet Union have the capability of destroying each other's military surveillance and communications satellites, as well as commercial satellites that could be used as backups. The U.S. Defense Department has attempted to harden its satellites against a Soviet antisatellite (ASAT) nuclear weapon with only limited success.

In the early 1960s, the United States tested its first ASAT nuclear device mounted on a Nike-Zeus missile. It was originally conceived to destroy warheads of the Soviet Fractional Orbital Bombardment System (FOBS), which could orbit Earth and then on command, drop down on its

(Courtesy of NASA)

FIG. 3-13. The Skylab space station.

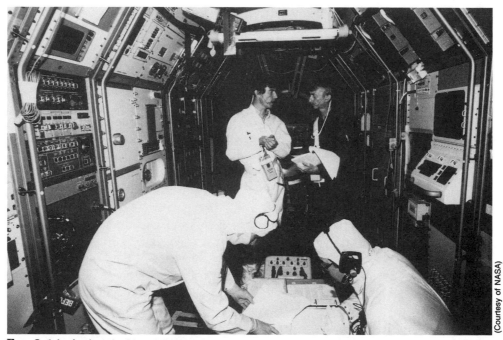

(Courtesy of NASA)

FIG. 3-14. A view inside of Spacelab 1 showing technicians conducting performance tests.

(Courtesy of U.S. Air Force)

FIG. 3-15. An artist's concept of an antisatellite missile launched from an F-15 aircraft.

target. When the ASAT was detonated in space, it sent out a powerful electromagnetic pulse (EMP) that affected satellites halfway around the world. Because of this problem, the 1967 Outer Space Treaty put an end to the basing of nuclear weapons in space and on the Moon.

Another idea was to maneuver a killer satellite close in, detonate it with conventional explosives, and penetrate the thin skin of the target satellite with shrapnel, destroying the delicate electronic components inside. The Soviets also have been reported to be testing powerful ground-based lasers used to blind the optics of surveillance satellites. Since the beginning of 1984, the United States has been developing small ASAT missiles mounted on F-15 aircraft (FIG. 3-15).

The major problem with the superpowers destroying each other's satellites is that this action can be highly destabilizing when tensions are on the rise, bringing nuclear war near at hand.

4

Detection from a Distance

REMOTE sensing is the ability to tell something about an object without being in direct contact with it. More specifically, it involves recording and analyzing electromagnetic (EM) energy from visible light, infrared radiation, microwave radiation, and all other forms of wave-propagated energy. In the past, remotely sensed data were derived primarily from visible light, using first our eyes and later cameras as sensors. However, visible light is only a very thin slice of the entire EM spectrum.

The reason we see only in this narrow band is that sunlight reaching the surface of the Earth is the strongest at these wavelengths. Certain properties of visible light allow the observer to collect a variety of information about the world around him. It is for this reason that the eyes of most animals, including man, function essentially in the same manner. Both eyes are separated so they view an object at slightly different angles, called *parallax* (FIG. 4-1). When the visual information is processed in the brain, certain spatial properties about the object can be discerned, such as its three-dimensional shape and its distance from the observer.

The color of an object is immensely valuable, and even the simplest of insects use the entire visible spectrum to locate brightly colored flowers for pollinating. Unfortunately, man's view of the world is extremely narrow in scope, more so than other species, and we are totally blind to important information conveyed in the upper and lower ranges of the visible spectrum. There is also a great deal of information about the universe contained in other wavelengths invisible to the eye.

THE VISIBLE SPECTRUM

The difference between the light we see and other parts of the EM spectrum lies in the wavelength, or energy (wavelength and energy are interrelated). Light waves are similar to waves

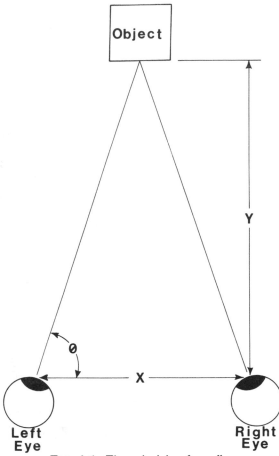

FIG. 4-1. The principle of parallax.

of near-infrared radiation between 700 and 1,500 nm and to a lesser degree, ultraviolet radiation from 250 to 400 nm. Because of the way our eyes are constructed, however, we simply cannot see these wavelengths.

The evolution of photoreceptors goes back over 2 billion years when primitive bacteria began using sunlight as a source of energy. Only a narrow region of the solar spectrum penetrates the atmosphere and reaches the ground. Therefore, these wavelengths have the greatest amount of energy (FIG. 4-2), and early organisms used this energy for photosynthesis. At first there was little distinction between plants and animals, and they both shared certain attributes. Eventually, diversity led to specialization, and organs were developed to perform specialized tasks, including sense organs.

The environment offers four basic properties that can be detected by sense organs: light, pressure, temperature, and a huge assortment of chemicals. These are translated into the five senses of sight, sound, touch, taste, and smell. Of these senses, only sight and touch involve the EM spectrum.* The first primitive eyes could only detect the difference between light and dark, and told the organism when it was day or night. As evolution progressed, animals acquired the ability to distinguish shapes and detect movement. Later, objects came into better focus, and the ability to distinguish colors evolved with more complex visual organs.

Some animals have the ability to see another world that we are totally blind to, just as dogs can hear sound frequencies much higher than we can. A large portion of the sensible world is closed to us, and our survival as a species has relied mainly on our ability to think, rather than on our acute senses.

An animal's eyesight is dependent on its environment, and fish living in murky water might

generated when a pebble is tossed into a calm pool. The distance between the peak of one wave and the peak of the next is the *wavelength*. Visible light wavelengths range from 400 to 700 nanometers (nm). A *nanometer* is one billionth of a meter, or 10 angstroms, which is another measurement of wavelength.

At the low end of the EM spectrum are radio waves, with wavelengths measured in meters, and at the high end are cosmic rays, with wavelengths only one ten-billionth of a meter. Our eyes perceive 400 nm as violet. As the wavelength increases, the light changes to blue, green, yellow, orange, and finally red at 700 nm. We are constantly being bombarded by other wavelengths in the visible spectrum, including a large amount

---

*You can feel sunlight on your skin as it is converted into thermal radiation.

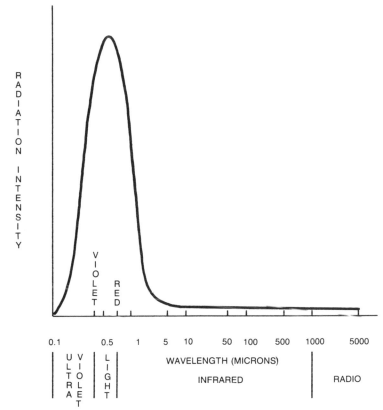

FIG. 4-2. The solar spectrum.

THE INFRARED SPECTRUM

only see reddish brown colors, while those living in deep, blue water have eyesights geared to the blue-green end of the spectrum, making their sighted world even more restricted than ours. Certain animals, like snakes, can sense beyond the visible spectrum into the infrared region, and can actually "see" heat radiating off a warm-blooded animal that might make a good meal. Honeybees and other insects can see the iridescent hues generated by ultraviolet light reflecting off the petals of flowers that make them highly attractive for pollination.

Insects have compound eyes, in which each facet has a complete optical system of its own. The insect looks around without having to move its eyes in the direction of the object it wants to observe. This is the same principle that is used by some electronic devices to open up a whole new world beyond our senses.

The infrared spectrum runs from the bottom of the visible spectrum, about 700 nm, to the top of the radio spectrum, around 100 micrometers (a micrometer, or *micron*, is one millionth of a meter). It is divided into two major regions: the visible infrared or *near-infrared*, and the thermal infrared, or *far-infrared*.

The heat we feel from the Sun is thermal infrared energy. It is generated by the conversion of visible light into thermal energy by absorption, which is mostly dependent on color. Light-colored objects reflect most of the light, while dark-colored objects absorb most of it. The Earth receives only about one-billionth of the Sun's total radiant output. Only about half of this amount ever reaches the ground, of which about one-third is reflected back into space (FIG. 4-3). The rest of the sunlight is reflected off cloud tops or is used to warm the atmosphere. The particles in the atmosphere also scatter much of the light, particularly blue light, which is the reason the sky appears this color.

About 90 percent of the solar radiation absorbed by the ocean is used to evaporate seawater to make clouds. When it rains, the clouds liberate trapped thermal energy, which then escapes into space. The Earth must reradiate into space exactly the same amount of energy it receives from the Sun, or the planet could become excessively hot or exceedingly cold, and life could not exist here.

Thermal infrared radiation manifests itself as waves of heat radiating off the ground on a hot day. It is responsible for mirages, which bend light rays to look like a pool of water in a dry desert.

Near-infrared radiation ranges from dark red, or 700 nm, to about 1,500 nm in the visible spectrum. This type of energy remains part of the visible region of the EM spectrum, even though it is not sensed by humans. Unlike thermal infrared radiation, near-infrared has nothing to do with heat, but instead is light with a longer wavelength.

Special photographic film enulsions are sensitive to infrared light, and objects that absorb near-infrared appear dark, while objects that reflect near-infrared appear light on infrared photographs. Vegetation such as conifers absorbs near-infrared, while broad-leaf vegetation reflects it, and this is useful for identifying certain types of plants. Infrared photography also can determine whether a plant is under stress from disease or drought, and indeed infrared film was originally developed for the military to detect camouflage using green plant or cut vegetation, which appears lighter than healthy foliage. Bodies of water have a high infrared absorption and usually register quite dark on infrared film. Photography taken with infrared film normally penetrates haze better than ordinary films, but will not penetrate dense haze or moist clouds.

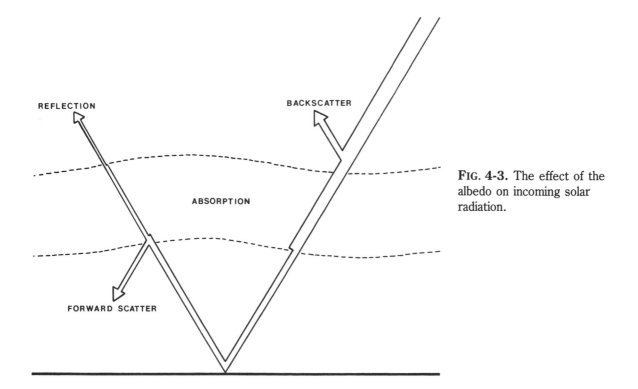

FIG. 4-3. The effect of the albedo on incoming solar radiation.

Multispectral sensors carried on board satellites (FIG. 4-4) also make use of the near-infrared wavelengths, which are generally broken down into various band widths that convey separate pieces of information about the environment on the ground.

The infrared spectrum contains a lot of information about the universe that cannot be obtained by other means. All stars, including the Sun, radiate enormous amounts of infrared radiation, but because the Earth's atmosphere absorbs a great deal of infrared, very little ever reaches the surface. Ground-based observations are possible in narrow windows in the spectrum where wavelengths of near-infrared radiation can penetrate the atmosphere. For far-infrared observations, telescopes have been carried aloft by balloons, rockets, and high-altitude aircraft. Practically the entire infrared sky has been surveyed by the Infrared Astronomical Satellite (FIG. 4-5), which operated throughout most of 1983. All objects above absolute zero (0 degrees

(Courtesy of NASA)

FIG. 4-4. The Thermatic Mapper carried by Landsat 4.

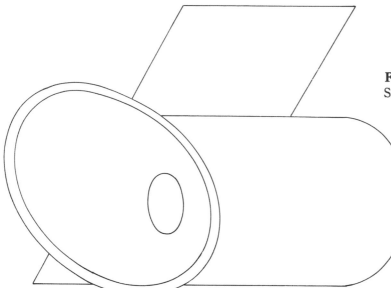

FIG. 4-5. The Infrared Astronomical Satellite.

Kelvin or −273 degrees Celsius) radiate infrared radiation, and many astronomical objects, which are principally made up of cold, dark material, radiate most of their energy in the infrared region. Diffuse radiation at infrared wavelengths has become a valuable source of information about the solar system, the galaxy, and the early universe.

THE RADIO SPECTRUM

Both visible and infrared sensors are severely limited by darkness and cloud cover because they must rely on sensing reflected sunlight or thermal emissions from the ground (FIG. 4-6), and the atmosphere absorbs or scatters much of this radiation. Radio waves, on the other hand, are not hampered by these conditions. The radio spectrum begins at the bottom of the infrared, around 100 microns, and extends all the way down to a frequency of a few *kilohertz*, 1000 cycles/second, in the very long wave band, which is used to signal submerged submarines. The radio spectrum is further divided into VHF, UHF and microwave frequencies.

Microwave remote-sensing devices employ active and passive systems, which can operate under the cover of darkness and in all weather conditions, although resolution might not be nearly as good as other sensing systems because of the relatively long wavelengths involved. Active sensors such as radar (FIG. 4-7) transmit microwave energy and observe the energy scattered back from an object. Passive sensors such as radiometers measure the electromagnetic energy emitted directly from objects. Because radar provides its own source of illumination, it can image the Earth uninterrupted, which is crucial to the observation of dynamic phenomena such as ocean currents, migrating ice floes, oil slicks, and changing patterns of vegetation (TABLE 4-1).

The brightness of every point in the radar image is determined by the intensity of the microwave energy scattered back to the radar receiver from corresponding points on the ground. The intensity of this *backscatter* is dependent on certain physical properties of the surface, such as slope, roughness, and vegetation cover. It is

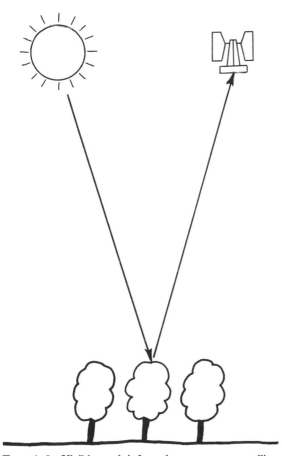

FIG. 4-6. Visible and infrared sensor on satellites require sunlight as a source of illumination and therefore are restricted to cloudless days.

also dependent on the electrical conductivity of the ground, which is directly related to soil porosity and moisture content.

Radar is capable of penetrating the thick foliage in rain forests to expose the ground. Microwave energy can reveal surface and near-surface geologic features such as faults, folds, and rock outcrops, as well as rock classification, which are important factors for mapping landforms. Earthquake faults disrupt radio-wave propagation near the point of rupture prior to a major earthquake, and there appears to be a direct relationship between the length of the disturbance and the magnitude of the earthquake. Radar altimetry via satellite—in which radio waves are bounced off the Earth's surface to precisely measure the height of the satellite from the surface, generally within a few inches—can determine the topography of the ocean floor by accurately measuring the height of the sea above certain terrain features. The sea level is influenced by the pull of gravity, which increases above undersea ridges and decreases over troughs (See Chapter 8.)

One of the major uses of radar is for tracking storm systems. Radar is capable of determining the location, size, movement, and intensity of storms. A network of weather radars throughout the United States monitors the formation and behavior of thunderstorms and other rain and snowstorms. The type of radar used by the National Weather Service (FIG. 4-8) can detect the presence of water drops and ice particles in the air. Radar also can be used to measure the intensity of the storm because the greater the size and concentration of precipitation particles, the more intense the radar echo.

Doppler radar, which operates similar to that used in highway speed traps, can measure the speed of the precipitation over the ground. Because tornadoes are characterized by their high wind speeds, Doppler radar is also well suited for their detection. In some cases, it is possible to detect evidence of tornado development up to 20 minutes before the funnel reaches the ground. High-powered radars are used to track hurricanes and typhoons when they reach within a couple hundred miles off the coast. The heading and speed of the hurricane is calculated to determine where and when the storm will make landfall, giving residents timely warnings to save them and their property.

Radio astronomy is the exploration of the celestial sky in the radio spectrum and has greatly aided in the classification of certain celestial bodies. Radio was first used in astronomy to accurately determine the distance to the Moon, Venus, and Mercury by bouncing radio waves off of them and measuring the time it took for the signals to return to Earth. The surface of Venus was mapped through its thick clouds using radars based on Earth and on the Pioneer Venus Orbiter.

Pulsars, which are rapidly spinning neutron stars, emit energy in the radio spectrum. When they were first discovered in 1967 by astronomers at Cambridge University using a very sensitive radio telescope, the regular pulses were thought to be sent by intelligent extraterrestrials. Since that time, radio telescopes have been pointed skyward, hoping to catch an alien transmission from one of our neighboring stars.

Man's first message to the stars was transmitted in November 1974 from the huge radio telescope at Arecibo, Puerto Rico. The message was sent in binary code, which is believed to be the universal language of computers, and any alien receiving this code should have little difficulty deciphering it. Unfortunately, radio signals from Earth might be too primitive for intergalactic communication—like jungle drums on an uncharted island.

THE ULTRAVIOLET SPECTRUM

On the upper side of the visible spectrum is *ultraviolet light*, which ranges from the top of the violet 400 nm to about 250 nm. Most of the ultraviolet radiation that reaches Earth is filtered out by the ozone layer in the upper stratosphere. The existence of the ozone layer is crucial because ultraviolet is one of the deadliest forms of solar radiation. Taken in small doses on the skin, ultraviolet light is essential for the production of Vitamin D, an important nutrient, and is also responsible for tanning the skin, the body's own protection against ultraviolet. However, overexposure to ultraviolet radiation can lead to skin cancer and *cataracts*, (a clouding of the lens of the eye and a leading cause of blindness). It

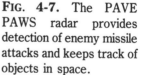

FIG. 4-7. The PAVE PAWS radar provides detection of enemy missile attacks and keeps track of objects in space.

(Courtesy of U.S. Air Force)

FIG. **4-8.(a)** The weather radar dome at National Weather Service station, North Little Rock, Arkansas.

FIG. **4-8.(b)** Meteorologist tracks storm system on weather radar.

TABLE 4-1. Albedo of Various Surfaces.

SURFACE	PERCENT REFLECTED
Clouds, stratus	
< 500 feet thick	25-63
500-1000 feet thick	45-75
1000-2000 feet thick	59-84
Average all types and thicknesses	50-55
Snow, fresh-fallen	80-90
Snow, old	45-70
White sand	30-60
Light soil (or desert)	25-30
Concrete	17-27
Plowed field, moist	14-17
Crops, green	5-25
Meadows, green	5-10
Forests, green	5-10
Dark soil	5-15
Road, blacktop	5-10
Water, depending upon Sun angle	5-60

also has harmful effects on plants and animals and exacerbates pollution problems, such as smog and acid rain.

Since 1979, satellites have measured a drop in the ozone layer of about ½ percent a year (FIG. 4-9). Over the Antarctic, a vast ozone hole appears in the atmosphere every spring (FIG. 4-10). The ozone depletion is blamed on the production of halo-carbons used in spray cans, refrigerators, foam plastics, and as industrial solvents. Unless this attack on the ozone layer is halted immediately, the consequences to the Earth's biota could be devastating.

For remote sensing of the Earth, ultraviolet light conveys very little useful information because so little of it ever reaches the surface. Insects make use of the ultraviolet spectrum to locate flowers for gathering nectar and for pollination. Some flowers reflect ultraviolet light, which gives them an iridescent appearance and makes them highly attractive to pollinators. All ultraviolet light is absorbed within the top few inches of water, so aquatic organisms have no use for it. Certain minerals will glow under an ultraviolet light source, called a *black light*, which makes it a useful tool for exploring geologic outcrops at night.

In orbit above the Earth's atmosphere, the International Ultraviolet Explorer satellite has studied the ultraviolet emissions from celestial objects like quasars, which resemble stars but emit power radio waves and are extremely distant. By analyzing the wavelength of a star's light, it can be determined whether a star is approaching or receding by its red shift. Using the Doppler effect, stars whose light is shifted to the red end of the spectrum are moving away from Earth and stars whose light is shifted to the blue end of the spectrum are moving toward Earth. Astronomers, studying the red shifts of the farthest galaxies, have been able to calculate the age of the universe, which is believed to be around 17 billion years old, give or take a couple billion years.

THE X-RAY SPECTRUM

Like ultraviolet radiation, most X-ray radiation, the next shortest wavelength, is also absorbed by the Earth's atmosphere. Substances heated to millions of degrees centigrade emit most of their energy in the form of X-rays. Large stars called *supernovas* are a major source of this radiation. When a supernova collapses, its outer shell is violently blown away and dispersed into space, where it heats the gases of the interstellar medium which, along with the collapsed stellar remnant left behind, emit copious amounts of X-ray radiation.

Solar X-rays were first detected by rocket-borne instruments in 1948 and were found to comprise only a millionth of the Sun's total output. Therefore, detecting X-rays from other stars required instruments with extreme sensitivity. The first satellite outfitted for X-ray observations in space was launched in 1970, and was succeeded by similar satellites that provided a wealth of

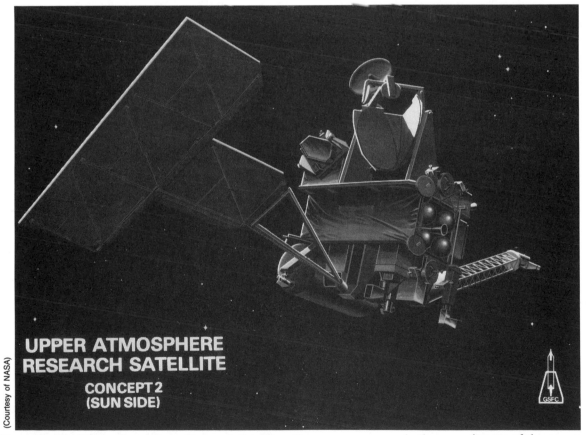

(Courtesy of NASA)

UPPER ATMOSPHERE RESEARCH SATELLITE CONCEPT 2 (SUN SIDE)

FIG. 4-9. The Upper Atmosphere Research Satellite will measure ozone and other constituents of the upper atmosphere on a global scale.

information about the X-ray sky. The X-rays were detected by Geiger counters with a thin window that blocked out visible and ultraviolet light but allowed X-rays to pass through. These devices had poor resolution, however, and increasing sensitivity meant increasing background noise from gamma and cosmic rays. With the development of X-ray optical systems used on board the Einstein X-Ray Telescope in the late 1970s, X-ray astronomy was able to match the sensitivity of optical and radio telescopes and has demonstrated the immense value of observing the heavens in the X-ray region of the EM spectrum.

THE COSMIC-RAY SPECTRUM

The most energetic radiation known are cosmic rays, which bombard Earth from every direction. These high-energy particles are far more powerful than anything physicists have created in particle accelerators. They have been studied since 1912 when the Austrian physicist Victor Hess made balloon flights with crude

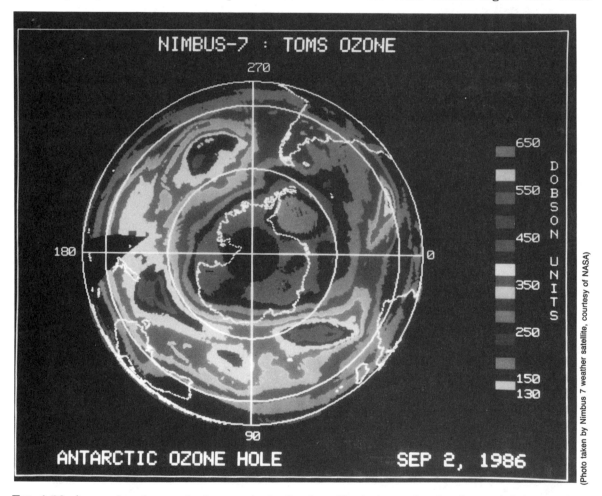

(Photo taken by Nimbus 7 weather satellite, courtesy of NASA)

FIG. 4-10. A map of total atmospheric ozone in the Southern Hemisphere, showing the ozone hole over Antarctica.

detectors to heights up to 16,000 feet. Hess found that the higher the altitude, the greater the intensity of these rays, which was evidence that cosmic rays came from outside Earth.

Cosmic rays are mostly hydrogen atoms stripped of their electrons, which make them high-energy protons. They spew out of collapsed supernovas along with X-rays. The protons are powerful enough to split atoms in the atmosphere, creating showers of elementary particles that survive to reach Earth's surface.

The aurora borealis and aurora australis (northern and southern lights) are caused by the emission of light from atoms excited by cosmic ray bombardment in the polar regions, where the magnetic lines of field converge.

Beginning in 1980, cosmic rays were detected on the ground by the Fly's Eye (FIG. 4-11), built and operated by the University of Utah. It is an array of 67 large metal drainage pipes with a 62-inch concave mirror mounted at one end and sensitive light detectors mounted at the other end. The cosmic rays streaking through the atmosphere excite nitrogen atoms in the air, causing them to fluoresce, and the bluish light is picked up by one of the detectors. Each cylinder is aimed at a different section of the night's sky, and the entire array is like the multifaceted compound eyes of an insect, keeping a constant vigil on the heavens.

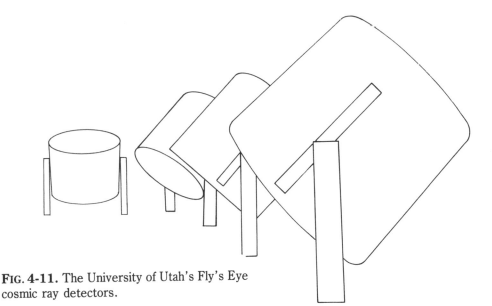

FIG. 4-11. The University of Utah's Fly's Eye cosmic ray detectors.

5

Image Interpretation

FOR the past three decades, advances in remote-sensing systems and computer technology have given scientists new tools for monitoring and understanding changes on the Earth. Remote sensing is widely used by biologists, geologists, geographers, agriculturalists, foresters, and engineers for evaluating natural and agricultural resources. Remote sensing also has been applied to assessing crop inventory and yield, forest pest damage and fires (FIG. 5-1), range condition; and mapping and monitoring vegetation, air and water quality, and other conditions important to the environment.

Through remote sensing, data is taken from satellites in orbit around the Earth, processed and enhanced by computers, and correlated with instruments on the ground. Ozone depletion, carbon dioxide buildup, and acid rain have dramatized the need for international cooperation,

for pollution does not recognize national borders. In order to deal with the problems, scientists must have accurate and up-to-date information provided by the most sophisticated and reliable instruments ever built by man.

Even though remote sensing offers the technology and a better perspective, ultimately it is the scientist who must provide the interpretation and society that must act on that interpretation.

SENSING THE ENVIRONMENT

When electromagnetic radiation encounters an object on the ground, it can be reflected, refracted, scattered, absorbed, or reemitted. If it is reflected or reemitted by matter, it must pass back through the atmosphere to a sensor on a satellite. However, because the atmosphere contains particulates (dust, soot, and aerosols), water vapor, carbon dioxide, and ozone, it can alter the intensity and composition of the radiation that eventually reaches the satellite sensor.

In most cases, the sensor is an electro-optical device that transforms EM radiation into electrical impulses. These impulses are converted by a computer into digital values, which are recorded on magnetic tape. Most electro-optical sensors are multispectral scanners (MSSs) that simultaneously record energy from several regions of the EM spectrum. The scanners can use a rapidly oscillating mirror to direct the radiation through an optical system, which separates it by filters into individual spectral bands (FIG. 5-2). The scanners also can employ a linear array of detectors that electronically scan as the satellite moves along in its orbit.

The Landsat MSS images in two visible bands and two near-infrared bands, and the Thermatic Mapper (TM) images in three visible bands and four near-infrared bands. Each band is then focused onto individual detectors with specific spectral sensitivities, and these detectors convert the radiation into electrical energy. The sensors scan the surface from west to east. The width of the scan swath (FIG. 5-3) is determined by the

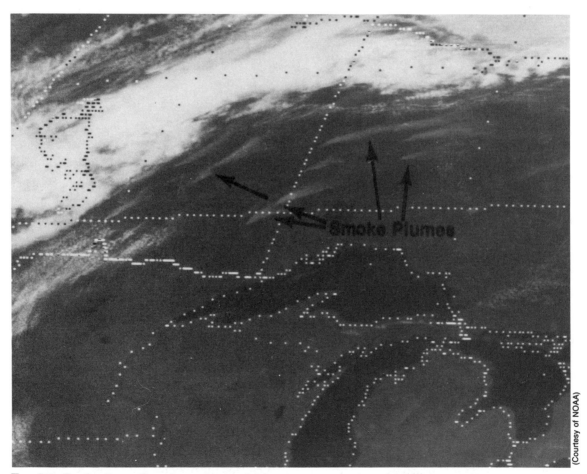

(Courtesy of NOAA)

FIG. 5-1. Smoke plumes from forest fires raging in remote areas of central Canada during the summer of 1976, consumed almost a million acres of woodland.

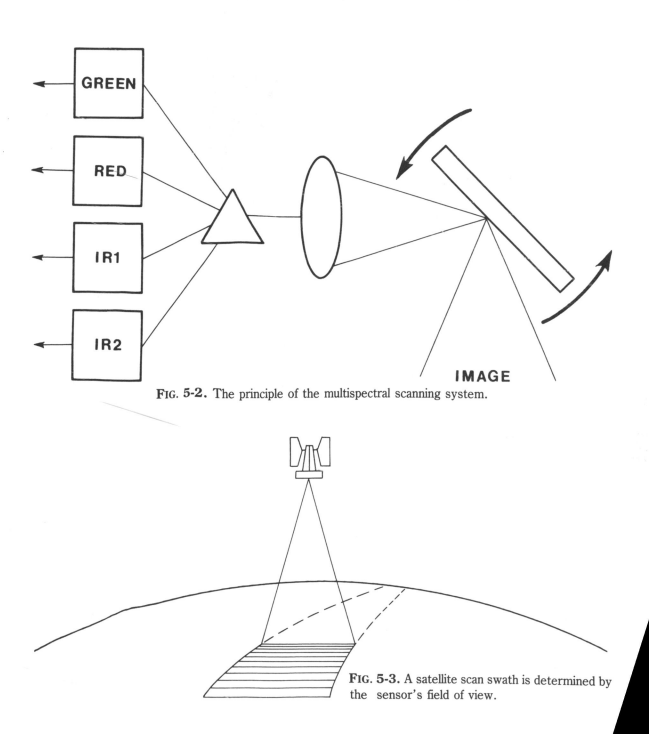

DETECTORS   PRISM   OPTICS   MIRROR

GREEN

RED

IR1

IR2

IMAGE

FIG. 5-2. The principle of the multispectral scanning system.

FIG. 5-3. A satellite scan swath is determined by the sensor's field of view.

sensor's field of view, which for Landsat is 115 miles and for SPOT is 37 miles.

The resolution of the system is determined by its ability to distinguish differences in surface information or its spatial qualities. The area on the ground represented by a given data value is called a *pixel*, a contraction of the words *picture element*. Pixels are like thousands of tiny rectangular pieces of a giant jigsaw puzzle. For Landsat 5, the resolution of each pixel at the *nadir*, the position directly below the satellite, is 270 feet for the MSS and 98 feet for the Thermatic Mapper. This might seem like a wide area, but keep in mind that it is viewed from a satellite over 400 miles above the Earth's surface and that the total field of view is 115 miles wide. Any improvement in resolution provides more detail, which in turn gives a better quality image.

The energy level of each individual pixel is recorded as digital values on magnetic tape. Each Landsat MSS pixel is assigned four digital values, one for each spectral band. The digital values for each band are then transmitted directly to receiving stations, located in many parts of the world. The raw data are not corrected for atmospheric and illumination effects, sensor calibrations, or distortions caused by the Earth's geometry, which introduces satellite altitude errors. Corrections are usually done at some later stage in the data processing.

To reconstruct the pixels into an image of a scene on the ground, the digital values are converted to corresponding shades of gray. These shades can be recorded on a photographic negative or displayed on a black-and-white or color video monitor. An image thus can be produced for each of the spectral bands or any combination of ratios or differences made between spectral bands. The final product can be made into a photographic print (FIG. 5-4), although the image represents data collected by an electro-optical system and not by a camera, and therefore technically it is not a photograph, but an image or a picture.

The advantage of collecting data in separate narrow spectral bands is that spectral-response patterns can be developed for diagnosing terrain features and changing resource conditions. The processing of digital spectral data involves identifying these patterns and maximizing the contrast between various classes or categories. The transformation of spectral data by calculating ratios or differences between spectral bands allows certain parameters to be *quantified*, or put into usable numbers that can be manipulated mathematically by a computer.

COMPUTER IMAGE PROCESSING

The computer is an essential tool for enhanced digital processing of images of the planets taken by spacecraft, especially images of Earth. An image comprises a prodigious amount of information, so most image processing requires large and powerful computers. In the past decade, a revolution in microchip technology has increased computer capacity and speed, and supercomputers like the Cray-1 and Cyber 205 can perform 100 million arithmetic operations per second.

Images from Landsat are recorded digitally in a form suitable for computer processing. Each pixel is assigned its own discrete numerical value in accordance to its position on the gray scale, a continuum of shades of gray ranging from white to black. An image that has been reduced to a set of binary numbers can be stored on magnetic tape or sent directly to the computer. The data then can be processed to obtain new sets of digital values, which in turn can generate a revised image. The image is usually displayed on a video monitor, where it is viewed directly or photographed. Higher quality photographs are made by taking the data from the computer pixel by pixel and projecting them directly on unexposed film.

After the data is in the computer, various mathematical operations can be carried out to enhance the visual quality of the image. Each operation can be done on any given point without reference to other points around it. An example would be adding a constant value of brightness to each pixel, similar to turning up the brightness control on a television set.

A mathematical operation called *image stretching* subtracts the minimum brightness value from each pixel and multiplies the brightness of each pixel by a constant, thereby increasing the contrast of the image and giving it more clarity. In a related process known as *false coloring*, a different color is assigned to each shade of gray in the digital image. As a result, areas that originally might have been distinguishable only by subtle shades of gray now have strikingly different colors, like comparing a black-and-white to a color photograph.

Another mathematical process compensates for light intensities that have been altered by passage through the atmosphere or through a poorly calibrated optical system, resulting in a

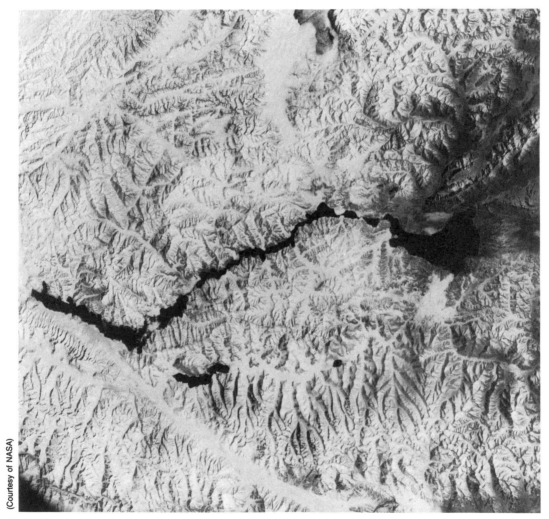

(Courtesy of NASA)

FIG. 5-4. The frontier between India and China in the Himalaya Mountain region, one of the most remote places on Earth.

degradation of the image. Yet another process performed on an image is similar to adjusting the tone controls on a stereo hi-fi set to obtain a better sound quality. In this case, however, treble and bass are replaced by reflectance and illumination, and these qualities are adjusted to obtain a sharper picture.

The most important application of digital image processing and enhancement is for satellite remote sensing of Earth. Landsat MSS records two images in the visible spectrum and two in the near-infrared spectrum. Areas of terrain that differ in certain physical properties reflect different amounts of radiation in each of the four spectral bands. Therefore, a forest yields a strong signal in the green band and a desert yields a strong signal in the red band. By combining images made at different wavelengths, it is possible to enhance the display. For example, a forest growing on partially open and rocky land will yield an image made with green light that shows intense reflections from the vegetation and an image made with red light that is sensitive to the bare ground. A computer can divide the brightness of each point in the green image by the brightness of the corresponding point in the red image, resulting in an image that can be ratioed to accentuate either the forest or the rocks. The technique is particularly suitable for identifying vegetation or mapping geologic features.

THE END USERS

Numerous private companies purchase computer-compatible tapes of Landsat data to perform individual custom computer processing and to extract information that normally does not show up in the standard processing of Landsat imagery. A single Landsat image views over 13,000 square miles of Earth's surface, and custom processing permits the detection and mapping of major regional structures associated with the development of entire geological provinces (FIG. 5-5).

Through special digital enhancements, it is even possible to map some of the more subtle surface expressions. To the geologist, structural features such as folds, faults, dips, and strikes of particular rock formations (FIG. 5-6), lineaments, landform and drainage patterns, and other anomalies are seen on Landsat imagery as possible traps for petroleum. The Landsat imagery cannot be used by itself to locate drill sites, but is integrated with other exploration tools such as seismic surveys and well logs.

Other natural resources also can be examined on Landsat images, including agriculture and forestry. New methods of Landsat interpretation are being researched and applied to many different disciplines by a variety of industrial, governmental, and academic concerns.

On September 27, 1985, in an effort to make Landsat imagery more available to private industry, the Earth Observation Satellite Company (EOSAT) was awarded a $250 million, ten-year contract by the U.S. federal government to take control of the Landsat program and develop the next generation of earth resource satellites. EOSAT is now the primary contact for marketing, ordering, and distributing of data from Landsat satellites. The company's customer service office is at the EROS Data Center, Sioux Falls, South Dakota. Prices vary depending on the imagery, but a 1:1,000,000-scale MSS color transparency costs about $150, a Thermatic Mapper color composite image sells for about $360, and a digital tape of a TM image runs about $3,300 at 1986 prices.

The intention is for EOSAT to make Landsat a viable commercial operation that will be used throughout the world in the development of renewable and nonrenewable resources. The company is not without competition from abroad, however, and the first of four French SPOT satellites, launched on February 22, 1986, has already returned tens of thousands of images of Earth. The satellite's 33-foot resolution, which is three times better than Landsat's Thermatic Mapper although each image covers a smaller area, guarantees SPOT a competitive place in the market.

EOSAT also serves as the marketing agent for the distribution of Landsat imagery from a new ground station at Beijing (Peking), China, which began operation in December 1986. The product line has been expanded to include *geocoded*, or map-oriented, Landsat Thermatic Mapper digital imagery which is overlaid with U.S. Geological Survey map projections. The geocoded images are useful to Geographic Information System users such as land-use planners, cartographers, surveyors, and state and local governments. Petroleum geologists are particularly interested in satellite data from China (FIG. 5-7), which had been unavailable prior to the construction of the new Chinese ground station.

Following *Landsat 3*, the decision was made to eliminate tape recorders on Landsat satellites, which previously stored acquired data and later

(Courtesy of NASA)

FIG. **5-5.** Thermatic Mapper image of Death Valley in California and Nevada revealing details of rock types and geologic structures.

transmitted it to a ground station when the satellite passed overhead. Instead, Landsat 4 and 5 use the Tracking and Data Relay Satellite (TDRS) system to acquire and transmit data. Unfortunately, many areas of the world like China were not covered by TDRS, and therefore no data was available for those geographical locations. Now, the new Chinese Landsat station receives images directly from the satellites as they pass over the country.

GROUND TRUTH

In order to make satellite imagery more meaningful, scientists need to test remote-sensing technology in the laboratory and in the field.

Biologists and other earth scientists design experiments on plants, including varying the water supply and nutrients, in order to find better ways to measure and understand ecological changes that are rapidly taking place on a global scale. They are trained to use new types of instruments (FIG. 5-8) and to interpret the data gathered by those instruments with the aid of advanced computer techniques.

A variety of sensors are evaluated for specific ecological applications. Direct identification of surface materials on a pixel basis is accomplished by sampling absorption features in the spectral band. The greater the number of spectral bands, the greater the amount of information that is

(Photo by J.R. Balsley, courtesy of USGS).

FIG. 5-6. Tilted sedimentary beds between Rawlins and Laramie, Wyoming.

conveyed in each pixel. Some surface materials have diagnostic absorption features only 20 to 40 nm wide. The Landsat scanners, which have bandwidths between 100 and 200 nm, are unable to resolve many of these spectral features.

The near-infrared spectra of vegetation contain information on plant pigmentation, leaf cellular structure, and leaf moisture content. When a plant is under stress from disease, pests, climate extremes, or mineral imbalances, its internal leaf structure changes. Although the changes might not be visible to the naked eye, they are visible to infrared scanners. Healthy vegetation shows nearly equal infrared intensities in two peak wavelengths (FIG. 5-9). Different plant species can be identified by their spectral ratio, which is computed by dividing the intensity of the first spectral peak by the intensity of the second and

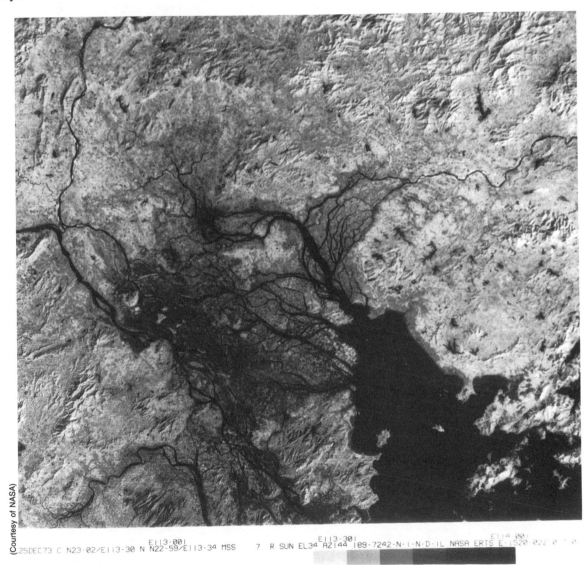

(Courtesy of NASA)

25DEC73 C N23-02/E113-30 N N22-59/E113-34 MSS   7 R SUN EL34 AZ144 189-7242-N-1-N-D-IL NASA ERTS E-1520-022

FIG. 5-7. A Landsat 1 image of the Canton, China area showing the Hei Chiang and Pearl Rivers.

multiplying the result by a constant. If the vegetation is under stress, one peak might be lower than the previous peak. When gathered in the appropriate format, this data permits early detection of disease, remote discrimination and mapping of plant species or communities, as well as the total amount of biomass or crop yield.

Most substances appear different when viewed at different wavelength bands and therefore have different spectral signatures. For most purposes, the subtle differences in surface characteristics become more evident when individual spectral band data are combined into a false-color composite. The colors are *false* because they represent spectral bands in the green, red, and near-infrared regions. Healthy vegetation appears bright red instead of green, urban areas appear blue or blue-gray, clear water appears black, and sediment-laden water is blue. Computers can assign any color to the spectral bands, so even though the image is still false color, vegetation can be made to appear green.

A process called *density slicing* can bring out contrasts in continuously changing values such as depths of water. Different false colors are assigned to certain ranges of values, resulting in an image that makes it easier for interpreters to spot essential differences.

Another process uses a target signature to locate particular surface features. The spectral signature of a desired feature such as a particular

(Photo by Tim McCabe, courtesy of USDA-Soil Conservation Service)

FIG. 5-8. An investigator uses a neutron probe to measure soil moisture in an avocado grove.

crop is identified, and the computer is instructed to highlight in false color all the areas in the image that have that signature.

Several factors can affect the observed pattern on the ground. Since 70 percent of the land surface is covered by vegetation, the most common environmental factors that influence target radiance are the quantity and orientation of vegetation and the amount of soil reflectance as compared with the vegetation reflectance.

Another factor is the physical geometry of the Sun and sensor in relation to the target. It can influence spectral characteristics used in computing vegetation indices for estimating the amount of vegetative growth, such as crop yields. The farther from the nadir, the greater the spectral difference. This factor is particularly important in MSS systems operating at relatively low altitudes. In such systems, the sensor view angle deviates rapidly from the nadir on either side of the flight line. Fortunately, computers can process these view-angle effects out of the data.

The solar angle, which is a function of latitude, time of year, and time of day, also can influence the observed spectral reflectance. Correcting for the solar angle greatly improves computer models developed from satellite imagery, enabling them to be applied over a wider range of latitudes and seasons for satellite remote sensing of the earth.

EROS DATA CENTER

On the basis of the potential of space technology and in response to the critical need for greater knowledge of the Earth and its resources, the U.S. Department of the Interior established the Earth Resources Observation System to gather and use remotely sensed data collected by satellite and aircraft of natural and man-made features on the Earth's surface. A large potential lies in the application of remote-sensing techniques for inventory and management of Earth resources and monitoring of the environment. Landsat imagery, by its broad area coverage, has identified previously unmapped geologic structures as targets for exploration of petroleum and minerals, and is being used to inventory water-impounded areas.

The repetitive coverage of satellite data provides information for land-use planning with a timeliness not previously possible. The capacity of detecting changes in land use has proven

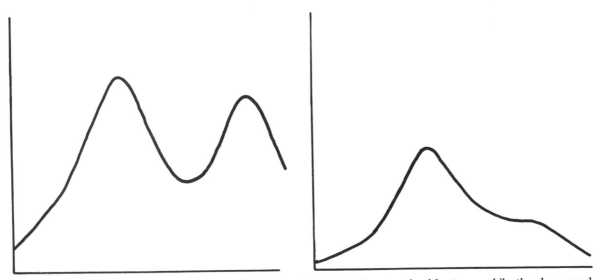

FIG. 5-9. The twin peaks in infrared intensity in the right graph denote a healthy tree, while the decreased intensity in the left graph indicates a dying tree.

effective in gauging environmental impact and assessing reclamation. The imagery also is used for evaluating range conditions over vast areas of the western United States.

The EROS Data Center is the national center for processing and disseminating spacecraft and aircraft-acquired photographic imagery and electronic data of the Earth's resources. It provides access to Landsat data, aerial photography, and other remotely sensed data acquired from research aircraft and spacecraft. At the heart of the data center is the central computer complex, which controls a database containing millions of images and photographs of the Earth's surface. It also performs searches of specific geographical areas of interest and serves as a management tool for the entire data-reproduction process.

The computerized data storage and retrieval system is based on a geographic system of latitude and longitude, and is supplemented by information about image quality, cloud cover, and type of data. The center can perform computer-assisted analysis of imagery. Digital analysis techniques are used to classify phenomena by their reflectance or emittance in different parts of the EM spectrum. The center also offers training sessions, stressing the use of data for particular applications such as agricultural inventory, water management, and natural resource management, which are vital for maintaining a healthy world environment.

6

Remote Sensing by Radar

RADAR has been put to a variety of tasks, including warning of enemy bomber and missile attacks (FIG. 6-1), keeping track of ships and planes, tracking weather systems, and catching speeding motorists. One of the most important uses for radar is imaging Earth from satellites and spacecraft. Radar's ability to see in the night and penetrate clouds makes it ideal for remote sensing, especially when it is necessary to keep a steady watch on constantly changing phenomena.

Spaceborne radar also can gather information about the surface of the Earth that mostly went undetected by other sensing devices. Radar's ability to penetrate dense vegetation has made it indispensable for mapping surface features that could not be reached by any other means. Structural and topographic features such as lineaments, folds and faults, stream drainage patterns, stratification, and surface roughness can be mapped by radar (FIG. 6-2). To a limited

extent, subsurface geological investigations can be carried out by radar for locating mineral deposits and oil traps. A great potential exists in monitoring ocean surface patterns for waves, currents, and large-scale eddies, which are important to the dispersion of ice floes, schools of fish, and oil spills. Radar altimetry from satellites has precisely measured the relief of the ocean surface throughout the entire planet, revealing the topography of the ocean floor and confirming the movement of tectonic plates and the drifting of continents.

SIDE-LOOKING RADAR

At the visible and infrared wavelengths, the atmosphere absorbs a large portion of the radiation even when the skies are clear. When the weather is cloudy and rainy, however, the performance of the sensors operating at these wavelengths is severely impaired. Therefore, an all-weather, day-or-night imaging system must be independent of

FIG. 6-1. Ballistic Missile Early Warning System radar at Thule, Greenland.

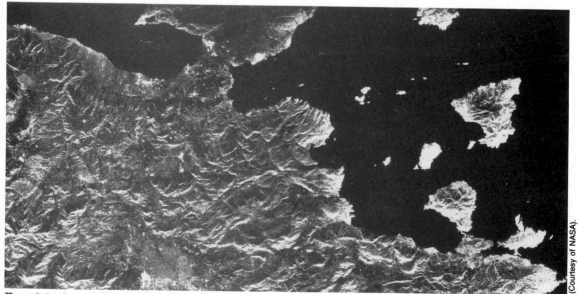

FIG. 6-2. Seasat Synthetic Aperture Radar imagery of geologic structure and topography near Knoxville, Tennessee.

solar energy and provide its own source of illumination. The radiation must be such that it is not attenuated or dispersed by water vapor. A radar imaging system operating at microwave wavelengths between 1 and 30 centimeters fulfills these requirements and more.

Although microwaves are about 100,000 times longer than the wavelengths of visible light, they are still short enough to provide resolution with enough detail for most geological and geographical purposes. A radar system operating at these wavelengths and looking out the side of an aircraft at the adjacent terrain can obtain images of the Earth's surface that display fine detail and spectacular relief. The reason for pointing the radar out the side of an aircraft or a spacecraft is to allow an increase in antenna size that is proportional to the resolution of the imaging system (a larger antenna provides greater detail).

The images made by radar and recorded on photographic film in many ways resemble an ordinary aerial photograph (FIG. 6-3), but there are also some fundamental differences. Because radar provides its own source of illumination and microwaves travel in straight lines, areas obscured by hills or other vertical structures are not illuminated and thus provide no return signal. As a result, the radar image contains shadows or voids, which are unlike areas poorly illuminated on a photograph by atmospheric scattering during low sun angles. Instead, the detailed character of the reflections on a radar image is determined by the wavelength and the polarization of the incident radiation and the geometry and electrical properties of the reflecting terrain.

A rough surface scatters the incident radiation in all directions, and only a small portion is returned to the radar antenna. A smooth surface reflects the incident radiation in only one direction and therefore acts like a mirror. If a smooth surface is perpendicular to the radar beam, the energy returned to the antenna is strong and the

(Courtesy of NASA)

FIG. 6-3. Shuttle Imaging Radar-A (SIR-A) image of Northern Peloponnesia and part of southern Greece.

image spot is bright. If, however, a smooth surface is at any other angle to the radar beam, none of the energy is returned and the image spot is dark. Therefore, cultivated fields become diffused reflectors, and clear-cut forests, lakes, and streams become *specular*, or mirrorlike, reflectors.

The radar system makes an image similar to the way the scan lines on a television picture tube make an image. Each point on the ground that reflects radar energy back to the antenna corresponds to a bright spot on a photographic film. A continuous image of the terrain is thus created line by line on the film. Relief in the terrain results because the portion of the microwave pulse reflected from elevated areas closer to the aircraft return to the antenna sooner than the energy reflected from the surrounding lower terrain (FIG. 6-4). In this manner, the radar image is similar to a photographic image with the images of elevated areas on the terrain displaced toward the aircraft's line of flight. The fan-shaped beam of microwave radiation points at right angles to the aircraft's line of flight, and all images of objects

on the ground are displaced toward the aircraft as a function of their height. Because the aircraft is moving slowly with respect to the speed of the microwave pulse, the dimension along the track is lacking in perspective and must be equalized with the dimension across the track on the final image in order to make a maplike radar survey of the Earth's surface.

The resolution of the radar image in the dimension across the track is determined by the *pulse length*, which is the length of time the radar transmitter is on during the propagation of each pulse of microwave energy. The signals reflected from two distinct objects can be resolved only if their respective ranges are separated by at least one-half the pulse length. A pulse length of $1/10$ microsecond (a *microsecond* is one-millionth of a second) covers a distance of about 100 feet and yields a resolution of 50 feet. If finer resolution is required, the pulse length must be reduced.

The resolution of the image in the dimension along the track is proportional to the radar beam width, which is determined by the length of the

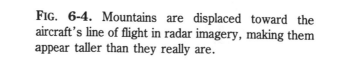

FIG. 6-4. Mountains are displaced toward the aircraft's line of flight in radar imagery, making them appear taller than they really are.

antenna. The longer the antenna is, the better the resolution. If two objects lie in the same direction along the track, giving them the same range, they can be resolved only if their separation is larger than the radar's beam width. Otherwise, they are in the radar beam simultaneously, the antenna will receive their reflections at the same time, and they will appear on the image as a single object.

Because the width of the radar beam spreads with range, objects farther from the antenna have less resolution than those closer. By taking advantage of the forward motion of the aircraft or spacecraft, it is possible to make a short radar antenna behave like a very long one, thereby making a synthetic antenna that is longer for more distant features and shorter for closer ones. This is called a *synthetic aperture radar* (SAR) and makes it possible to obtain high-resolution images of terrain at great distances away.

RADAR IMAGES FROM SPACE

One of the most dramatic uses of radar on orbiting spacecraft was for the purpose of mapping Venus, whose dense cloud layers make it impossible to observe the surface of the planet by any other means except by radar imagery made directly from Earth. However, this method is limited to the middle and lower latitudes on the side of Venus facing Earth when the two planets are close together. Radars on board the Pioneer Venus Orbiter and Venera 15 and 16 spacecraft have returned images of volcanoes, mountains, valleys, craters, and other large-scale features on the planet's surface (FIG. 6-5). The radar imagery reveals terrain features on Venus that are much like those that might have existed on Earth during its early years. Venus also has a complex geology including global tectonics. Parallel bands of high radar reflectivity appear to be crust folded and

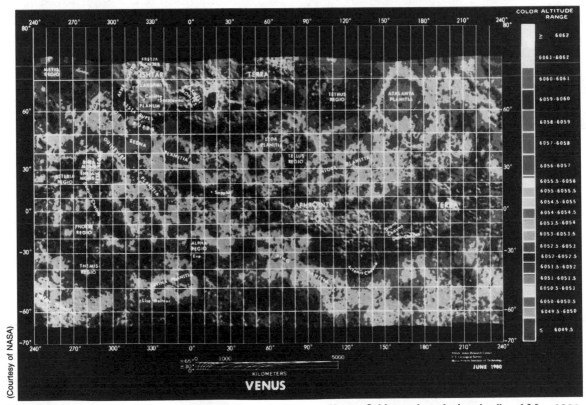

FIG. 6-5. Radar image of Venus landforms from the Pioneer-Venus Orbiter taken during April and May 1980.

faulted into mountains by horizontal compression, similar to the way the Appalachian Mountains formed by the collision of North America and Africa.

The Moon also was imaged by a synthetic aperture radar system carried by the *Apollo 17* spacecraft. The SAR system made pictures of the lunar surface from an altitude of about 60 miles with a resolution of about 30 feet. However, the surface of the Moon is never obscured by clouds like the surface of Venus or certain parts of Earth.

Placing radars on orbiting spacecraft presents greater problems than those encountered with low-flying aircraft because resolution diminishes as the altitude of the sensor increases. At an altitude of about 150 miles above the Earth's surface, the radar would have to have an antenna over a mile long to achieve the same resolution as optical systems.

The solution to this problem was the same as that found for side-looking airborne radar, which is the use of the synthetic aperture principle. As the side-looking orbital radar system traverses along its flight path, it transmits a beam of microwave radiation at an oblique angle to the Earth's surface. The signal is backscattered by objects on the ground, received by the antenna,

and recorded. Those signals detected at various points along the flight path are later combined by a data processing system to form a two-dimensional image of the terrain. In this manner, the moving radar antenna functions like an elongated antenna whose maximum length equals the distance along the track for which the target is within the beam of the antenna. The resolution of SAR is independent of altitude and is upwards of three times greater than the resolution achieved by Landsat sensors, which rely on reflected sunlight in the visible and near-infrared ranges.

Radar images of Earth from space are being used in such fields as geology, meteorology, oceanography, and the study of renewable resources. Radar imagery is particularly well adapted for mapping landforms (FIG. 6-6). Many surface and near-surface geologic features such as folds, faults, outcrops, and other structures are associated with variations in the appearance of the surface, especially in its topography, surface texture, or vegetation cover. Erosional processes also generate recognizable features, such as stream drainage patterns. Because the backscatter of microwave radiation is very sensitive to changes in the physical properties of the surface, these features and patterns are clearly

(Courtesy of NASA)

FIG. 6-6. Shuttle Imaging Radar-B image of Mount Shasta in the Cascade Range of northern California.

observed in the radar imagery. In some cases, radar actually enhances topography and brings out more clarity. The best details are observed when the incident radiation is almost perpendicular to the direction of the topographic trend.

Unlike visible and infrared sensors, which depend on the position of the Sun, the angle of incidence on radar sensors can be controlled, providing greater flexibility. Therefore, radar sensors are ideal for detecting subtle structural features that influence the topography.

Radar is particularly helpful in studying heavily forested regions like those in the tropics where the topography cannot be revealed by other means. In arid parts of the world, it is possible to penetrate several feet below the surface to uncover the geological and archaeological history of the region. Radar information can discriminate between different rock types, which is helpful in mapping lithological units over a wide area. The radar data is even more important in vegetated areas where an image of the surface cannot be made by Landsat or other multispectral sensing systems.

Radar signals are strongly reflected by man-made structures because flat perpendicular surfaces tend to form corner reflectors that return a major portion of the incident radiation and metal structures act like antennas that strongly reradiate the incident energy. As a result, many man-made structures are visible in radar images even though they are smaller than the resolution of the system. Combined with Landsat imagery, radar images can improve the capability of remotely monitoring urban development on a large scale.

The intensity of radar backscatter is also highly sensitive to ocean waves, and any ocean feature that alters the roughness of the surface can be detected by satellite radar. The velocity of surface currents is determined by surface winds and affected by the bottom topography in shallow areas. Therefore, coastal bottom features have been observed in radar images even though microwave radiation does not penetrate the water to any significant depth. Internal waves, eddies, rings (or *gyres*), thermal fronts, and suspended sediment also affect the surface roughness of the sea and are observed in radar images. In addition to surface roughness, surface motion can be detected and the velocity of currents can be determined.

Ice features such as ice floes, ice ridges, and open channel ways can be distinguished in radar imagery (FIG. 6-7), making it possible to monitor the motion of the polar ice, which in turn is driven by the polar currents. In addition, the continuous monitoring of icebergs is vital to the shipping industry and offshore drilling rigs in polar waters.

ARCHAEOLOGY BY RADAR

A large area of ancient civilizations in a region of Central America known as *Mesoamerica*

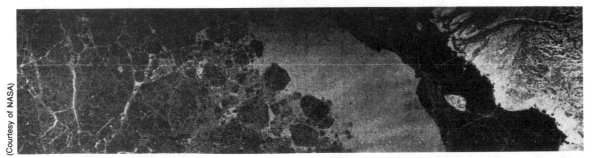

FIG. 6-7. Seasat Synthetic Aperture Radar image of pack ice in the Beaufort Sea northwest of Alaska.

contains the remains of many cultural traditions that existed during the Classical period between A.D. 250 and 900. However, many areas that were once occupied by these peoples, who were mostly Mayan, are covered by dense vegetation, especially in the lowlands, which receive abundant rainfall. The heavy foliage has hampered rapid survey techniques such as aerial photography and wide-range ground reconnaissance, and this in turn has given archaeologists an incorrect perspective about the size and distribution of ancient cities.

There are presently over 300 Maya centers, ranging in size from a small paved courtyard and a few buildings to the largest known site, Tikal in northern Guatemala, with 85 courtyards and hundreds of buildings, which once housed a population of 50,000 people. The Mayans used a modified form of slash-and-burn agriculture, in which fields were cleared by burning off the vegetation and then cultivated. In some areas, the Mayans were supported entirely by intensive agriculture, and extensive swamps in the lowlands might have been drained by canal systems for cultivation. Such a system of widespread agriculture could have supported large populations, perhaps as many as 1,000 people per square mile. Then around A.D. 900, for unknown reasons, the entire Mayan civilization collapsed.

In order to improve survey data on the Mayan lowlands, scientists searched for a remote-sensing tool that was both rapid and reliable. What was needed was a sensor system that could penetrate foliage, silt, and root cover to map ancient roads, causeways, and other man-made structures. Airborne side-looking radar is such a system. The archaeological sites appear on the radar imagery as one or more large, irregular bright spots, which indicate large buildings, or as distinct conical shadows cast by large mounds.

Regular grid patterns in the radar image represent ancient canals, but only the largest and most widely spaced canals are picked up by the radar. The grids are closely associated with known areas of swamps, lakes, ponds, and watercourses.

In addition, the canals could have been used as a means to transport goods, which also explains how large populations in cities like Tikal were supplied. Raised fields that were once cultivated lands lie between the canals and produce a gray line pattern, which is created by a slight difference in elevation of the vegetation. The radar data indicate that the Mayans intensively cultivated the southern lowlands by draining swamps, which were the most productive land available. This explains the location of large cultural centers on the edges of large swamps. The use of such an intensive and sophisticated system of agriculture requiring considerable management might also explain in part the demise of the Mayan civilization.

In November 1981, the Space Shuttle Imaging Radar-A (SIR-A) using a synthetic aperture imaging radar system returned startling pictures of the Sahara desert (FIG. 6-8) that force geologists to change their conception of the desert's underlying structure. Lying beneath the desert sands are a vast network of river valleys as wide as the Nile valley, along with smaller river channels, gravel terraces, desert basins, and bedrock structures. Also scattered in the sand are Stone-Age human artifacts which indicated that humans and life-sustaining water sources were once present in an area that is now uninhabitable.

As long as the sands are extremely dry, subsurface penetration of several feet is possible with radar signals of certain wavelengths. Investigators at the sites searching through the sediments discovered one of the last great river systems on Earth, and its activity was much more intense than had been previously thought. Channels hundreds of thousands of years old twisted their way through valleys that were millions of years old. The radar images also revealed that water was trapped in the deep valleys, which has important implications for hydrological exploration in the region.

When the scientists dug into the ancient river banks, they discovered dozens of artifacts, including 250,000-year-old stone axes that resembled those found on the surface. The axes were made of volcanic material, which meant they could not be dated by the potassium-argon method and are much too old for carbon-14 dating, so the margin of error could be as much as 100,000 years either way. The artifacts appeared to be from campsites where people known as *Homo erectus* lived and made tools about quarter of a million years ago. The evidence indicated that the filled-in valleys under the sand might once have been roadways for early humans migrating out of Africa into Europe. This evidence suggests that an earlier hypothesis that *Homo erectus* gathered in small groups around desert springs is incorrect.

(Photo by E.D. McKee, courtesy of USGS)

FIG. 6-8. A Skylab photograph of linear dunes in the northwest Sahara desert of northern Africa.

It is also possible that sediments buried deep in the valleys date back several millions of years and contain human artifacts of that time period. Unfortunately, the sediments are deep, probably 1,000 feet below the surface, making excavation too expensive. What is significant, however, is that radar can provide a new tool for archaeological exploration in the Sahara Desert, as well as other deserts of the world.

HURRICANE HUNTING WITH RADAR

Hurricanes are the greatest storms on Earth, and over the years, a considerable amount of effort has been devoted to studying their dynamics and developing models for forecasting their intensity and travel path. Once a hurricane makes landfall, its surface winds cause the most damage, either directly against structures or indirectly from a storm surge where the wind piles up water in front of the hurricane and causes flooding and beach erosion. It is the dynamics of these surface winds that are of paramount interest to researchers and forecasters alike.

Methods of estimating hurricane-force winds, which were previously based on the state of the sea, often have limited accuracy. The only reliable

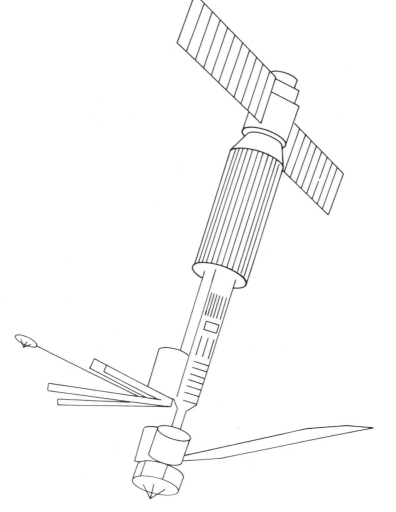

FIG. 6-9. The Seasat satellite.

method of obtaining the surface wind speed was by flying through the hurricane with specially equipped aircraft called *hurricane hunters*. However, this method is extremely hazardous, and a better method using radar remote-sensing techniques might be more reliable and safer.

The surface roughness of the ocean correlates directly with the wind speed, with greater winds producing larger waves. Radar backscatter from the sea can be used to measure both the wind speed and wind direction. The first radar measurements of a hurricane were obtained from *Skylab*, and the same concept was used in the design of the Seasat Synthetic Aperture Radar (FIG. 6-9), which was placed into orbit in June 1978. In addition, radar-equipped aircraft were able to obtain accurate data while flying at higher altitudes with considerably less risk (FIG. 6-10).

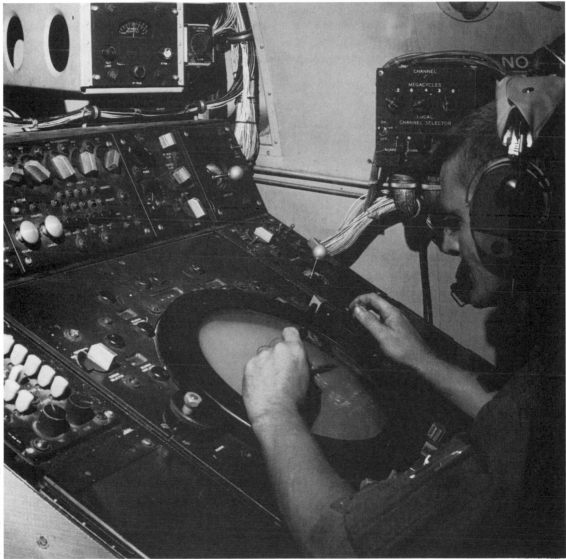

FIG. 6-10. A hurricane hunter crewmember plots storm on aircraft radarscope.

Other oceanographic applications for which radar is well suited are the all-weather, day-or-night surveillance of winds, surface waves, currents, weather fronts, and changes in surface temperature. Each of these weather phenomena affect the surface roughness of the ocean, which can be detected by radar satellites. Surface waves are visible on the radar image as periodic regular changes in the image tone, caused by the backscattering of radar energy off the flanks of swells. Internal waves are observed on the radar imagery as a series of convex strips, whose crests can range up to several miles long and are separated by dark regions. Some of the waves overlap, causing distinctive interference patterns.

Surface features associated with the Gulf Stream in the North Atlantic Ocean were observed on radar images. The boundary at the western edge of the Gulf Stream corresponded to an abrupt change in the image tone, along with numerous parallel streaks, giving the impression of flowing water.

The monitoring of the polar ice cover is another major application of spaceborne radar. The delineation and extent of the ice cover is important for determining the heat flow from the ocean to the atmosphere, which has a major effect on the climate the world over.

7

Watching
the Weather

CLOUDS and the rain they bring might spoil a family outing, but they also play an important role in transporting heat and moisture around the globe. Tropical storms and monsoons distribute much needed rainfall throughout the world. Anomalous conditions such as unusual warming in the South Pacific can have a drastic effect on these storm systems, sending unusual weather around the world. In addition, clouds dramatically affect the surface temperature by reflecting radiant energy from the Sun and trapping infrared energy emitted by the Earth, causing temperatures to lower substantially on cloudy days and cloudless nights. Approximately 70 percent of the energy Earth receives from the Sun goes to drive the climate system, and most of the rest is lost to space.

At any one time, clouds cover half the surface of the Earth (FIG. 7-1), but they account for about two-thirds of the planet's *albedo*, or reflectivity. Clouds therefore aid in regulating the temperature

of Earth. If global temperatures become too hot, more water is evaporated from the oceans to form clouds, which in turn block out more sunlight. If global temperatures become too cold, the opposite condition results. It is estimated that the second half of this century has been more cloudy than the first half, possibly because of air pollution, contrails from jet aircraft, and a southward shift of the polar front, which could instigate more clouds and storms. The main reason for putting meteorological satellites into orbit around Earth is to keep a weather eye out for clouds and the storms that accompany them.

FORECASTING THE WEATHER

Hurricanes, tornadoes, floods, and other weather-related catastrophes continue to take an inordinate number of lives and cause thousands of injuries every year despite advances in technology and skill in forecasting and warning. The cost to federal, state, and local governments in the United States is over $3.5 billion annually. During the widespread tornado outbreaks of 1974 and 1984 (FIG. 7-2), the Weather Service warnings, the broadcast of warnings by the news media, and the rescue operations by community officials and agencies were responsible for saving thousands of lives.

Over the last two decades, there has been a substantial increase in population along the Atlantic and Gulf coastal areas. People along the coast have witnessed many near misses and the associated weather effects of hurricanes. Yet because of timely warnings, more than 40 million people along the hurricane-vulnerable coasts have never experienced the devastation of a major hurricane (TABLE 7-1).

Floods and flash floods are more widespread than any other natural hazard and are the number one stormy weather killer in the United States. Deaths from flash floods are now approaching 200 each year, compared to an average of less than 70 a year during the preceding 30-year period. Damages from flash floods are now nearly ten times what they were in the 1940s. Despite these hazards, people continue to build on the floodplains, and nearly 85 percent of all federal expenditures for disaster relief is flood related.

The modern-day American meteorologist has at his disposal a network of satellite and radar images, computer projections, and up-to-the-minute reports from points around the country. His primary responsibility is giving timely notices of life-threatening weather phenomena such as hurricanes, tornadoes, thunderstorms, floods, and blizzards. The meteorologist must know in what area a hurricane is likely to make landfall, must

(Courtesy of U.S. Air Force)

FIG. 7-1. A view of the Earth showing extensive cloud cover.

keep track of thunderstorm development (FIG. 7-3), and must estimate the amount of moisture that will end up as runoff and produce flooding. Forecasting has improved significantly over the past three decades and bears no resemblance to the archaic methods of the last century. Many of today's forecasting tools and methods came out of military research during World War II. Despite the miracles of our technological age, however, weathermen, much to their chagrin, still make mistakes. Generally though, the forecasters' track record is fairly good, running at about 90 percent accuracy for 24-hour predictions. With each added day into the future, the accuracy drops another 10 to 15 percent, until by the fourth or fifth day, the forecast becomes little more than an educated guess. Nevertheless, today's 5-day forecasts have the same accuracy as 2-day predictions 20 years ago.

There are over 300 National Weather Service facilities operated by the National Oceanic and Atmospheric Administration (NOAA) in the United States and elsewhere. In a single year, over 20 million weather observations are processed by the Weather Service from all over the world, and approximately 2 million forecasts and warnings are issued. Meteorological data are collected from the land, the sea, and the upper atmosphere by people from many countries.

In the course of a typical day, the National Meteorological Center, the nerve center of the National Weather Service, located in Camp Springs, Maryland, receives approximately 50,000 surface reports from land, 3,000 reports from

FIG. 7-2. Approximate areas affected by the 1974 super outbreak (solid line) and 1984 Carolina (dashed line) tornado outbreaks.

TABLE 7-1. Chronology of Major U.S. Hurricanes.

DATE	AREA OR HURRICANE	DAMAGE IN $ MILLIONS	DEATH TOLL
1881	Georgia and South Carolina		700
1893	Louisiana		2,000
1893	South Carolina		1,000-2,000
1900	Galveston		10,000
1913	Great Lakes		250
1919	Florida Keys and Texas		600-900
1928	Okeechobee		1,800
1935	Florida Keys		400
1938	New England	$ 300	600
1944	Atlantic Coast		400
1954	Carol	450	
1955	Diane	800	
1957	Audrey, Louisiana		400
1960	Donna	400	
1961	Carla	400	
1965	Betsy	1,400	
1969	Camille	1,400	250
1970	Celia	450	
1972	Agnes	2,100	
1975	Eloise	500	
1979	Frederic	2,300	5
1979	Claudette	400	
1979	David	300	
1983	Alicia	675	
1985	Elena	550	

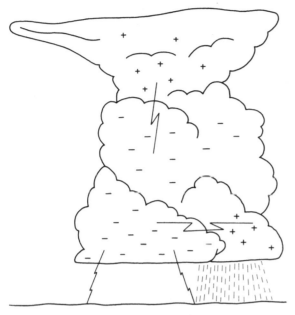

FIG. 7-3. The structure of a thunderstorm.

ships, 4,000 upper air observations, and 3,000 to 4,000 reports from aircraft. Additionally, satellite data, including cloud photographs, atmospheric temperature soundings, and sea surface temperatures, are sent to many receiving stations on the ground. Ocean buoys provide additional information on sea temperatures, ocean currents, and air-sea interactions.

NOAA also operates the National Severe Storms Forecast Center in Kansas City, Missouri. It is responsible for forecasting severe local storms, especially tornadoes. The responsibility for forecasting the path and intensity of hurricanes, other tropical disturbances, and associated sea conditions is divided among three Hurricane Forecast Centers, located in Miami, Florida; San Francisco, California; and Honolulu, Hawaii. The National Meteorological Center also provides assistance on locating and tracking the eyes of dangerous hurricanes.

The National Weather Service is limited in its ability to provide industry and other private concerns with specialized weather forecasts tai-

lored to individual needs. Therefore, many industries that depend on the weather for their livelihood and have millions of dollars at stake hire their own meteorologist or use the services of a meteorological consultant. Most major cities across the country have radio and television stations that employ the services of a nongovernmental weather forecaster. Many large airlines maintain their own organizations to collect, distribute, and interpret information about the weather.

The private meteorologists subscribe to the same meteorological data provided by the National Weather Service. The major differences are the interpretation of the data for specific areas such as ski resorts or orange groves. The services of a skilled company or consulting meteorologist could maximize benefits when the weather is favorable and minimize losses when it is unfavorable.

SATELLITE WEATHER

The first polar-orbiting TIROS weather satellites were launched in the early 1960s and were little more than television cameras sent into space to observe clouds, the harbingers of storms. The pictures showed cloud details, which were useful for understanding weather patterns.

The early satellites were especially helpful in tracking storms forming over tropical seas, where weather data were sparse or nonexistent. When the satellite passed over one of these weather systems, it relayed the pictures back to a ground station and alerted meteorologists to the storm's position. Armed with this information, the forecasters placed these systems on weather maps, estimated their direction and intensity, and provided the necessary warnings.

The sporadic use of satellite data continued until 1965, when new weather satellites resulted in images of cloud patterns on a worldwide basis. Thus, for the first time, all storms could be seen at least once daily, ensuring that no tropical cyclones, or hurricanes would approach land without warning.

Continuous coverage of the weather over an entire region was provided by the Applications Technology Satellite (ATS-1), launched into geostationary orbit in 1966. In the succeeding years, higher resolution visible imagery and infrared sensors provided coverage both day and night. Meteorologists were able to obtain satellite imagery at 30-minute intervals. When these images were viewed in sequence, they created a "motion picture" of weather developments, which allowed forecasters to improve their predictions of local severe storms.

In the latest meteorological satellites, the television cameras have been joined by multispectral sensors in the infrared and microwave regions. These imaging capabilities have widened the scope of applications from simply the ability to take pictures of clouds to uses in oceanography, including imagery of the Gulf Stream, biologically rich upwelling currents, and areas of ice cover. Temperature soundings through the atmosphere, which were once the domain of weather balloons, are now performed routinely by weather satellites, although they do not have nearly the same degree of accuracy. The real advantage of satellite temperature soundings is their use in thinly populated areas and the oceans, where conventional observations are sparse or nonexistent.

In addition to atmospheric temperature, the latest geostationary satellites of the GOES series can image water vapor, liquid water, and ice clouds. Because dry air is generally associated with descending motion or high pressure and moist air is generally associated with ascending motion or low pressure, the water-vapor imagery can be used to infer dynamic features of the atmosphere's circulation (FIG. 7-4). The water-vapor imagery also can resolve sharp boundaries of moisture in the midlatitudes, indicating the boundaries of the upper atmosphere jet stream (FIG. 7-5), which plays an important role in initiating and steering weather systems. Satellite imagery also can identify frontal systems, developing thunderstorms, dust storms areas of rain and snow, and fogbound areas, and can determine the ground temperature.

On land, weather balloons are launched from weather stations (FIG. 7-6) to obtain information about the winds aloft. However, 70 percent of Earth is covered by oceans, and these areas do not have routine balloon measurements. Satellite imagery can provide comprehensive observation of the motion of the atmosphere by measuring cloud displacements. By tracking clouds, which move with the wind, it is possible to describe the motions of the atmosphere at cloud level over most of the Earth. The height of the clouds is determined by measuring their temperature observed in infrared imagery and comparing it with a vertical sounding of the temperature versus height. In areas devoid of clouds, wind estimates can be provided by tracking water-vapor features in water-vapor radiation imagery from GOES satellites.

The cloud and vapor-drift wind data are not as accurate as those derived from balloons because of the imprecise tracking of the cloud features, limitations in satellite data quality, and uncertainty in height measurements. The last item generally represents the largest of the errors. The wind data derived by geostationary satellites operated by the United States, the European Space Agency, Japan, and India are fed into global weather-computer forecast models.

Satellite imagery and sounding data also can fill in the gaps between routine ground observations and upper air balloon soundings over a wide area of the United States. This is particularly important for understanding and predicting thunderstorms. The patterns of clouds in satellite imagery portray dynamic and thermodynamic processes in the atmosphere (FIG. 7-7). When satellite information is combined with radar and surface and upper-air observations, many of the important processes in storm development and evolution are better understood, which in turn can lead to better forecasts of severe weather.

Many outbreaks of severe weather are generally associated with squall lines, which tend to form in areas of converging air or fronts. Viewed from a satellite, a frontal system (FIG. 7-8) forms a distinct band of clouds generally associated with the jet stream and moving in a roughly west to east direction. Thunderstorms that develop along the frontal system can produce damaging hail and tornadoes. Squall line developments of this sort are routinely observed

(Courtesy of NASA)

FIG. **7-4.** A space shuttle photograph of large cumulonimbus clouds building up in Zarie, central Africa.

in GOES imagery even before the development of thunderheads and before their detection by radar. This ability to locate the early development of severe storms along squall lines and track their motion has led to timely warnings of impending danger.

The observation of tropical cyclones since weather satellites were first launched has provided meteorologists with a wealth of information about the relationship between cloud patterns and their position, motion, and intensity. The zed spiral structure and eye formation of these intense well-organized spiral structure and eye formation of these intense storms which are highly distinctive in satellite imagery (FIG. 7-9) bear a close relation to the high wind speed associated with them.

Certain cloud features such as the appearance of the storm's eye, its banding, coiling of the curved cloud band, and the size of the evolving cloud pattern are used to estimate storm strength.

Because of the ever-changing nature of the clouds that form the pattern, however, these essential features are not always clear-cut and can take on a variety of appearances at various levels of intensity. Infrared imagery provides continuous measurements of the cloud top temperature, and enhanced infrared and digital infrared data make the analysis of storm intensity simpler and more objective than imagery in the visible range.

Microwave temperature observations of the upper troposphere are used to determine the intensity of the cyclone's warm core, which is related to the storm's intensity. The satellite cloud imagery is also used to forecast tropical cyclone motion based on time-lapse motion pictures taken at frame rates of roughly every 30 minutes. In addition, steering wind currents are estimated from satellite imagery to forecast the motion of the tropical cyclone and estimate its place and time of arrival should the storm make landfall.

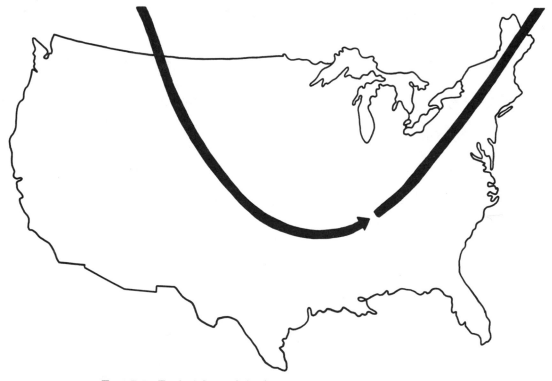

FIG. 7-5. Typical flow of the jet stream over the United States.

NUMERICAL FORECASTING

The histories of modern weather forecasting and modern computing have been intertwined since World War II. Weather modeling is ideal for a computer because the behavior of the atmosphere obeys certain complex but well-defined equations of fluid motion. The effects of air, water, mountains, and coastlines are translated into mathematical equations that can be manipulated by a computer. The computer model contains equations that act like wind, clouds, and rain around the entire globe, making it an imitation of real life. In order to make a forecast, programmers set the model running in the computer and observe what happens.

As long as there are accurate data and fast and powerful computers, meteorologists can attempt to stay ahead of the weather. Weather observations worldwide are more plentiful than ever thanks mainly to satellites. Reports from a network of aircraft, ships, and buoys supplement weather stations on land. Vast uncovered areas remain, however, particularly in the Southern Hemisphere which is predominantly ocean. In addition, the atmosphere is so unstable and chaotic that even small errors grow rapidly. For this reason, only coarse computer simulations of the weather can be provided, which is limited to predictions of about two weeks in advance.

The primary use of cloud data from meteorological satellites has been for weather forecasting, but there was also a need for a quantitative interpretation of the imagery to be fed into computer models. It was generally believed that numerical weather predictions were seriously hampered by insufficient data and that satellites offered the only timely and economical means for describing the atmospheric circulation in remote areas, especially the large ocean areas of the Southern Hemisphere and the North Pacific. Low-level winds were inferred from their action on the ocean surface, and upper-level winds were estimated from cloud motion and time-lapse imagery of evolving cloud patterns. Models of the

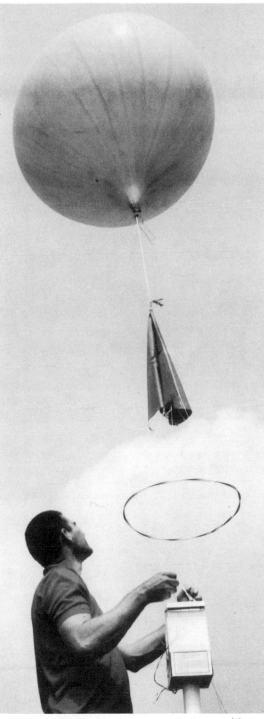

(Courtesy of NOAA)

FIG. 7-6. Launching of weather balloon with radiosonde.

evolution of subtropical storm systems provided information on the barometric pressure during those storms. Relative humidity was modeled by the moisture definition of cloud imagery.

With the addition of satellite soundings of temperature and moisture, more quantitative methods were developed to determine the state of the atmosphere in all parts of the world all at once. In theory, improvement of this initial state of the atmosphere should enable the models to better predict the future state of the atmosphere.

The complex mathematical models of the atmosphere are run by the fastest computers in the world: the supercomputers. The National Meteorological Center in Camp Springs, Maryland, uses a Cyber 205 for its global forecasts. The European Center for Medium-Range Forecasts at Reading, England, employs meteorologists from 17 countries and uses the largest and fastest supercomputer, the Cray X-MP, to produce global forecasts three to ten days in advance.

The atmospheric model is a very refined mathematical model that is constantly changing and includes the formation and dissipation of clouds, rain and snow precipitation, and changes in sea

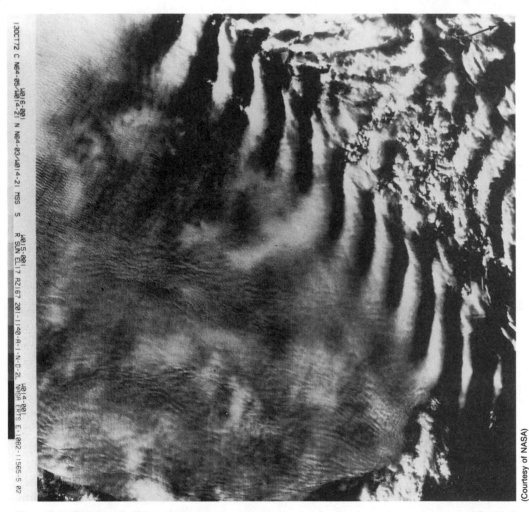

(Courtesy of NASA)

FIG. 7-7. Landsat 1 view of a unique wave pattern in clouds along the east coast of Iceland.

C-1. This remarkable photograph of Hurricane Gladys about 150 miles SW of Tampa, Fla., was taken on October 18, 1968, from the Apollo 7 spacecraft at an altitude of 97 nautical miles. Winds speed was 80 knots at the time. (NASA)

C-2. The sunglint on the Coral Sea to the northeast of Australia clearly shows the dynamics of the ocean surface, its currents, and eddies. This photo was taken aboard the Space Shuttle *Challenger* on September 16, 1983. (NASA)

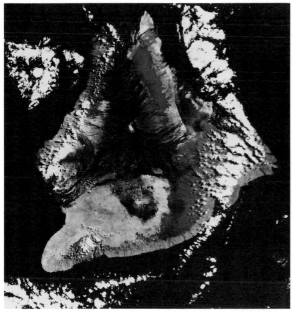

C-3. This false-color mosaic of two ERTS pictures covering the island of Hawaii, gives some indication of the youth of the island's volcanoes in the vivid dark colors of the lava flows radiating from the summit of Mauna Loa (north). (NASA)

C-4. The small Henderson Island peeks through the South Pacific clouds. Astronauts aboard the Space Shuttle *Challenger* were asked to find and photograph the island using the best available survey map, which was from the 1800s. (NASA)

C-5. NASA's STS-41-C astronauts used a 70mm handheld Hasselblad camera to record a prime example of the effects of continental drift on this planet. These tectonic forces within the Earth cause various surface features to move slowly in various directions. In this east-looking view, portions of the countries of Yemen, People's Democratic Republic of Yemen, Ethiopia, Somali, and Djbouti are seen. The Red Sea is at lower left and the Gulf of Aden at center. Differential movement has brought about the formation of the seas. At Djbouti, the internal forces are now splitting the earth and permitting more and more water to flow through a system of fissures into sub-sea-level basins in Ethiopia. Geologists predict that the basins will "soon" (in geologic terms) become open bays of sea. This frame was one of the visuals used by the 41-C astronauts in their April 24, 1984, postflight press conference. (NASA)

C-6. The Mauna Loa volcano (alt. 13,018 feet) on the island of Hawaii can be seen in this 70mm frame photographed through the overhead windows of the Earth-orbiting Space Shuttle *Challenger*. The close look reveals lava flows from the active volcano. (NASA)

C-7. Iran has a great desert in the east called the Dasht-E-Lut. An area known as Namak-Zar, about 100 miles east of Kerman, is at the center of this remarkable photograph taken December 9, 1981, from the Space Shuttle *Columbia*. A series of very long, parallel ridges and depressions are eroded in the desert dry lake bed sediments. They are elongated in the direction of the prevailing winds. This is one of the world's great wind erosion features and is called a *yardang*. At the left edge of the photo is a great sand sea deposit of large sand dunes, which are oriented at right angles to the wind. (NASA)

C-8. Straight of Gibraltar and western Mediterranean photographed in sunglint from the Space Shuttle *Challenger* on November 15, 1984. The Sun's reflection on the water delineates surface texture, which corresponds to the ocean's dynamics. The large group of waves connecting Gibraltar with the Moroccan coast is the response of the ocean to a tidal pulse moving into the Mediterranean. The low-level wind shear line extending out from the Moroccan coastline into the Mediterranean for approximately 40 miles is the southern extremity of the wind funnelling through the straight. This is the first high-resolution photograph ever taken of this phenomenon, permitting detailed delineation of individual waves which comprise the wave packet. (NASA)

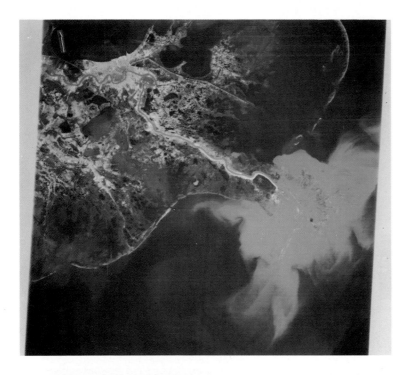

C-9. This color composite photo from the Landsat-1 satellite was taken from an altitude of 914 kilometers (568 statute miles) of the Mississippi Gulf Coast area on January 16, 1973. These colors—green, red and infrared—seen and recorded separately by the satellite were combined at NASA's Goddard Space Flight Center, Greenbelt, Maryland. Healthy crops, trees, and other green plants, which are very bright in the infrared but invisible to the naked eye, are shown as bright red. Suburban areas with sparse vegetation appear as light pink, and barren lands as light gray. Cities and industrial areas show as green or dark gray, and clear water is completely black. The cloud of mud and silt coloring the water around the river mouth graphically illustrates the process of delta building. (NASA)

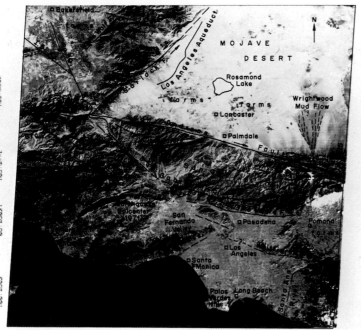

C-10. The San Andreas Fault is clearly visible in this Landsat-1 picture with Los Angeles, California, prominent in the foreground. The indentation running from one side of the picture to the other about midway on the right and running off the photo to the upper left is the San Andreas Fault. The Mojave Desert with the dry bed of Rosamond Lake is upper right and Palmdale is just below it. The fault line contrasts starkly with the San Gabriel Mountains rising at the edge of the fault. The Wright Wood Mud Flow, the fan-shaped area on the right edge of the middle of the picture, also provides contrast with the desert. (NASA)

surface temperature. It takes into account the global geography, especially the effects of mountains. Measurements of the atmosphere throughout the world come in continuously over a global telecommunications network and are fed into the Cray computer, which produces a forecast four times daily. The Cyber computer at Camp Springs receives about 100,000 separate measurements and produces forecasts up to five days in advance twice daily and longer forecasts once a day. However the global models are too coarse and cover too large an area to be of much use for local predictions. Furthermore, despite the huge number of observations that go into every computer prediction, the atmosphere cannot be measured precisely all at once.

(Courtesy of NOAA)

FIG. 7-8.(a) GOES weather satellite view of frontal clouds in the Great Lakes Region.

FIG. 7-8.(b) Formation of frontal storms.

CLIMATE MODELING

The climate is always changing: last year's climate was different from this year's; last decade's climate was different from this decade's; last century's climate was different from this century's; the climate 18,000 years ago at the height of the last ice age was different than it is in this age, and the climate 100 million years ago when the dinosaurs roamed throughout every corner of the world was considerably different than it is today.

In the future, the climate will continue to evolve. Part of that evolution will be driven by natural forces as it was in the past, and part of it will be driven by an entirely new force—that created by man's activities. We might already be or will soon be feeling the climatic effects of air pollution, particularly carbon dioxide, particulate matter, acid rain (which destroys trees), and halocarbons (which degrade the ozone layer). In order to prepare for such an uncertain climatic future, climatologists simulate the climate

1830 12SE84 38A-1 02526 24893 DIANA

(Courtesy of NOAA)

FIG. 7-9. Hurricane Diana off the North Carolina coast.

worldwide with the aid of computers. Using mathematical equations to represent the interaction of ocean, atmosphere, and land, the computers can calculate how the climate will evolve in accordance with physical laws.

The mathematical climate models cannot simulate all the complexities of the real climate. They can only reveal the consequences of certain actions, such as the effect of a major volcanic eruption. This capability also can be applied to past climates, for example to learn what the climate was like in the Cretaceous period during the reign of the dinosaurs.

Some of the processes that influence the climate—the advance and retreat of glaciers, the global spreading or destruction of forests, the motions of lithologic plates, and the transfer of heat from the surface to the deeper layers of the ocean—operate very slowly. A model designed to forecast next week's weather can ignore these variables, and they are treated as just ongoing conditions. However, a model designed to simulate the climate in the distant past or far into the future must include all these variables and more. The most sophisticated climate models, which deal with the general circulation of the atmosphere, predict the evolution of temperature, humidity, wind speed and direction, soil moisture, and other important climatic variables.

Even the most complex model is limited in the amount of detail it can offer because, at present, no computer is powerful enough to calculate all climatic variables around the world quickly enough. Instead, calculations are done on a three-dimensional grid system with widely spaced points on the order of a couple hundred miles wide and twenty miles deep. Unfortunately, many important weather phenomena like clouds are smaller than individual grids, and predicting changes in cloudiness is an essential part of climate simulation.

PROBING THE ATMOSPHERE WITH LASERS

Remote sensing of atmospheric properties by laser is known as *lidar*, an acronym for *li*ght *d*etection *a*nd *r*anging, and is analogous to radar. A short laser pulse is transmitted through the atmosphere, and a portion of the radiation is reflected from atmospheric constituents such as air molecules, clouds, dust, and aerosols. The interaction of the incident laser energy with these substances produces changes in the intensity and wavelength, depending on their concentration. As a result, information on the composition and physical state of the atmosphere can be deduced by analyzing the lidar data. In addition, the distance of the substances that interact with the laser beams can be determined by the delay of the backscattered radiation, similar to the way radar works.

Some of the most important uses of lidar are the measurement of the movement and concentration of air pollution near urban areas, the chemical emissions from industrial plants, and other chemicals in trace amounts in the atmosphere. Lidar has been used to measure the speed and direction of winds near storms, especially at airports where aircraft might be in danger from wind shear generated by thunderstorms.

Lidar is also used to track the global circulation of volcanic ash emitted into the atmosphere by recent volcanic eruptions, like the 1982 eruption of El Chichon in Mexico (FIG. 7-10), considered the dirtiest volcano in the past 100 years. The climatic effects were felt the world over, and had it not been for remote sensors, including those carried on satellites, the effects of volcanoes on the climate might have gone by unnoticed.

8

Observing the Ocean

THE world would benefit significantly from better climate predictions because climate variability affects our lives and a large part of our economy. Improved predictions depend on a better understanding of the effects of the ocean on the climate. New space technology can provide the necessary global synoptic descriptions of the upper ocean (those that exist simultaneously over a broad area) and its interaction with the atmosphere.

In order to understand this air-sea interaction process, ocean currents, eddies, surface winds, and radiation must be measured simultaneously worldwide. Such measurements can only be obtained by sensors on satellites and complemented by measurements taken on the surface of the oceans. This information is then fed into computers to help analyze and model the data.

In addition, satellite measurements of global biological processes, especially the near-surface marine ecology, are an essential key to understanding the biogeochemical cycles of the ocean. Ocean color measurements can provide a determination of the chlorophyll content of the surface layers which relates to the primary production of the sea. High-precision satellite radar altimeter measurements of the topography of the ocean, which is affected by winds, currents, and gravity, can provide a synoptic global description of ocean circulation. The simultaneous measurements of currents and winds will give investigators a better understanding of the physics of large-scale and long-term air-sea interactions, which is central to climate prediction for the protection and control of the environment.

SEA SURFACE TEMPERATURE

Taking measurements of the ocean by ships and buoys is time-consuming and expensive. In addition, vast areas of the ocean cannot be observed in this manner. As a result, there is a great need to make continuous global measurements of the ocean, and only remote sensing from space can accomplish this task. Over the last two decades, NASA has developed spacecraft with a wide range of sensors for taking ocean measurements.

One of the most important ocean measurements is sea surface temperature. The NOAA polar-orbiting weather satellites (FIG. 8-1) provide data on sea surface temperatures, making two million global observations monthly. Precision has improved over the past several years with the use of multiple infrared bandwidths on the NOAA advanced very high resolution radiometer (AVHRR). When corrected for water vapor and aerosols in the atmosphere, the accuracy of the data is quite good—comparable to data obtain from drifting buoys. These buoys transmit data on their position and sea surface temperature to NOAA polar-orbiting satellites, and the data is used to ensure the accuracy of satellite-obtained sea surface temperatures.

When the Mexican volcano El Chichon erupted in April 1982, the atmospheric dust and aerosols, which blocked radiation from the earth, had a marked effect on the satellite's ability to monitor sea surface temperature. The discrepancy between satellite-derived data and data taken from the surface of the sea was useful for monitoring the evolution and dispersion of the volcanic dust cloud.

Sea surface temperatures observed by NOAA polar-orbiting satellites also can be used to monitor surface currents. A relatively cool tongue of water appears annually in the eastern equatorial Pacific. It is a result of the equatorial upwelling of cold, deep water and the westward-flowing southern equatorial current. Because the equatorial currents and upwelling are driven by the wind, the sea surface temperature patterns taken from satellites can be used to verify predicted patterns in computer models of the ocean circulation.

Although atmospheric models for weather prediction have been in use for several decades, global ocean circulation models have only recently been used to predict ocean response patterns.

(Courtesy of NASA)

FIG. 8-1. The TIROS-N/NOAA weather satellite provides the capabilities for mapping ozone, determining the Earth's radiation budget, and performing search and rescue missions.

The sea surface temperature patterns associated with equatorial long waves have been compared in both satellite measurements and computer models. These long waves form north of the equator and spread westward at a speed of about one knot. During an *El Niño event*, which is an unusual warming of the South Pacific occurring about once every five to eight years, the westward-blowing trade winds relax, and the upwelling and the south equatorial current weaken. As a result, the long-wave patterns are no longer present in satellite-derived measurements of the sea surface temperature. This pattern could be useful for predicting El Niños, which can wreak havoc on global weather patterns.

The El Niño event of 1982–83 caused major physical and biological changes in the equatorial Pacific. The greatest anomalies were near the eastern Pacific coasts, with notable changes in surface and subsurface temperature, sea level, surface currents, and *phytoplankton* (microscopic marine plant life). During the entire course of the El Niño, a systematic sampling of the ocean was carried out using a variety of sensors on orbiting satellites. These sensors provided extensive and synoptic coverage with resolutions of one square mile or better and a precision sufficient to detect major changes in sea surface temperatures, coastal upwelling intensity, and ocean circulation patterns. Along the coast of southern California, small patches of relatively cool water were less intense—a sign of weakened coastal upwelling.

Along the coasts of Ecuador and Peru, the northward-flowing Peru current normally carries cold water rich in nutrients that support large fisheries in the region. During an El Niño, a tongue of warm water spreads along the equator across the Pacific (FIG. 8-2) and prevents the upwelling of cold water from below, which in turn increases the sea surface temperatures upwards of 10 degrees Fahrenheit and more above normal. Although the Peru current itself is not significantly

FIG. 8-2. Extent of El Niño warming in the Pacific.

weakened by this action, the upwelling water is warm and nutrient poor. This condition devastated the Peruvian anchovy industry during the El Niño of 1972–73, and it has not yet recovered.

Practically every continent was struck by weather-related disasters during 1982–83. Africa, Australia, and Indonesia were plagued with droughts, dust storms, and brush fires. Over a million people faced possible famine because of drought-related crop losses. In Australia, thousands of starving, thirsty cattle and sheep had to be destroyed and buried in mass graves. At the other extreme, areas in the central and equatorial Pacific received copious amounts of rainfall, which disrupted the economy as well as the ecology. The worst devastation to the ecology occurred when 17 million birds on Christmas Island disappeared without a trace. Peru and Ecuador were hit with the heaviest rain in recorded history, and some rivers were forced to carry a thousand times their normal flow. The island of Tahiti, which had not seen a typhoon in 50 years, was hit by six that season. In the North Atlantic, where hurricanes prowl during the late summer and fall, only two were reported—the fewest number of hurricanes in 50 years. Major storms marched in procession across the United States, causing severe flooding on the West Coast and in the South and forcing tens of thousands to leave their homes. The event was blamed for the deaths of upwards of 2,000 people and $8 billion in proper-

ty damages worldwide. It was one of the most disastrous climatic events in modern history.

MARINE PRIMARY PRODUCTION

Small, simple organisms called phytoplankton are responsible for more than 95 percent of all marine photosynthesis. They are the primary producers in the ocean and occupy a key position in the marine food chain. They also produce 80 percent of the breathable oxygen, as well as regulate carbon dioxide. Because of this crucial role that phytoplankton play in the marine ecology, which occupies 70 percent of the Earth's surface and because the distribution of phytoplankton is a key parameter in evaluating changing environmental conditions on a global scale it is essential that accurate methods of assessing global marine primary production are developed (TABLE 8-1).

At present, satellite remote sensing of ocean color is the only tool that can provide information on marine primary production worldwide. The surface waters of the ocean vary markedly in color, depending on suspended matter such as phytoplankton, silt, and pollutants. Some parts of the ocean are a beautiful azure blue because they contain no phytoplankton or sediment, and the color is determined by the optical properties of the seawater alone. By comparison, the waters of the North Atlantic are green because they are rich in phytoplankton. The organisms contain chlorophyll, which uses primarily blue light to manu-

TABLE 8-1. Productivity of the Oceans.

LOCATION	PRIMARY PRODUCTION (TONS PER YEAR OF ORGANIC CARBON)	PERCENT	TOTAL AVAILABLE FISH (TONS PER YEAR OF FRESH FISH)	PERCENT
Oceanic	16.3 billion	81.5	.16 million	0.07
Coastal Seas	3.6 billion	18.0	120.00 million	49.97
Upwelling Areas	0.1 billion	0.5	120.00 million	49.97
Total	20.0 billion		240.16 million	

facture carbon compounds for their growth. Because of this pigmentation, phytoplankton scatter light mostly in the green and infrared regions, just as do terrestrial plants.

The amount of phytoplankton, or *biomass*, is determined by analyzing the differential absorption and reflection of various wavelengths of light. The relative reflectance of blue wavelengths to yellow-green wavelengths is inversely proportional to the concentration of phytoplankton biomass. In the open ocean where the biomass is low, the water has a characteristic deep blue color. In the temperate coastal regions where the biomass is high, the water has a characteristic greenish color. In waters with high chlorophyll concentrations, blue light is so strongly absorbed that very little leaves the ocean. Also, in regions of high near-surface concentrations of chlorophyll, significant quantities of red light can be emitted from the sea by chlorophyll fluorescence.

The radiance values used for calculating chlorophyll concentrations are corrected for atmospheric contamination. Light penetrating the sea surface is backscattered by phytoplankton in the water, retransmitted back into the air, and diffused by the atmosphere before it reaches the satellite sensor. Therefore, these atmospheric components must be removed from the signal for the phytoplankton biomass to be accurately determined. Sun glint and suspended sediment load also make it difficult to determine phytoplankton biomass. Although chlorophyll cannot be accurately determined in waters with heavy concentrations of suspended sediments, the imagery is useful in assessing sediment load and in tracking river plumes (FIG. 8-3).

Dramatic changes in satellite measurements of chlorophyll concentrations around the Galapagos Islands off the Pacific coast of Ecuador were associated with the unusual oceanographic conditions during the 1982–83 El Niño. The redistribution of food resources might have contributed to the reproductive failure of seabirds and marine mammals on the islands during this time. The ocean current patterns around the islands are complex and greatly influenced by the equatorial undercurrent, a subsurface, eastward-flowing current about 600 feet thick. During a period when the sea surface temperatures were anomalously high, satellite imagery documented a major redistribution of phytoplankton around the Galapagos Islands. The patterns of chlorophyll distribution were drastically altered, and streamers of phytoplankton-rich waters were observed to extend northeastward away from the islands. Compared to normal conditions, there was a threefold reduction in chlorophyll concentrations and a twentyfold decrease in primary production. This decrease was felt on up the food chain and brought about a deterioration in the physical condition of predators, including seabirds, marine mammals, and fish.

Under normal conditions, ocean color measurements of chlorophyll are useful in locating good fishing grounds (FIG. 8-4) because marine life generally concentrate in ocean frontal zones and eddies that tend to concentrate phytoplankton. The color patterns are also valuable indicators of ocean currents.

OCEAN CIRCULATION

The dynamic nature of the ocean is reflected in almost all the features on its surface. Radar backscatter is totally controlled by the small-scale surface topography, including the short gravity and capillary waves and the local tilt of the surface, which is a result of the presence of large waves and swells.

Surface waves are visible on the satellite radar imagery as periodic changes in the image tone. Internal waves are observed on radar imagery as a result of their effect on the surface roughness. The large currents associated with these waves modify the surface waves overlying the oscillations by sweeping surface oils and materials together to form a smooth strip or by enhancing surface roughness from surface current stress. On the radar imagery, the smooth regions appear darker and the rough areas appear brighter than the normal sea surface. The internal waves are usually observed in packets or groups consisting of a series of bright, narrow convex strips separated by dark regions, with the length of each crest ranging upwards of several tens of miles.

Some waves can interact with each other and produce interference patterns. Surface features that are associated with major ocean currents (FIG. 8-5) appear on radar imagery as an abrupt change in image tone or numerous streaks pointing in the general direction of current flow. The change in image tone might result from the water motion and temperature changes, which affect surface roughness and consequently radar backscatter.

Eddies play an important role in mixing the ocean, just as tropical cyclones help mix the atmosphere (FIG. 8-6). They are swirling pools of chemically distinctive water, several tens to hundreds of miles across and can reach depths of three miles. Their rotation is clockwise, or

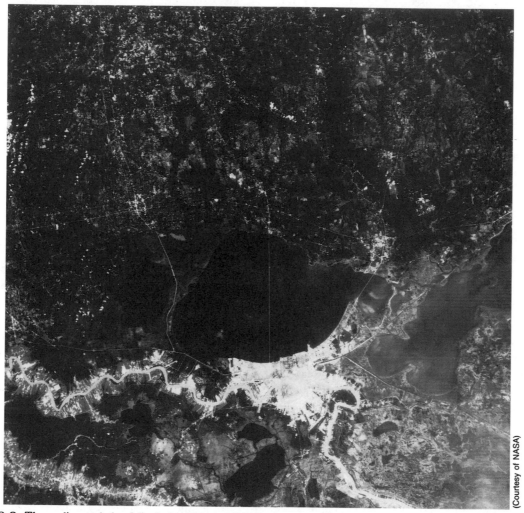

(Courtesy of NASA)

FIG. 8-3. The sediment-laden Mississippi River winds its way through the Mississippi Delta southeast of New Orleans, Louisiana where it discharges into the Gulf of Mexico.

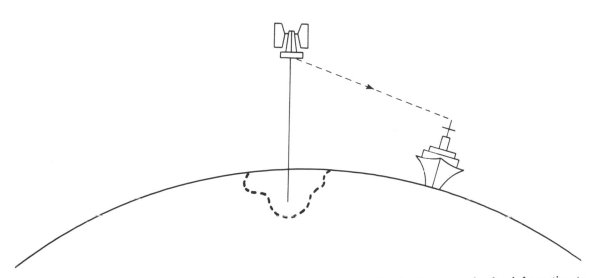

FIG. 8-4.(a) Fishing from space. Satellites measure primary production in the sea and relay information to fishing trawlers.

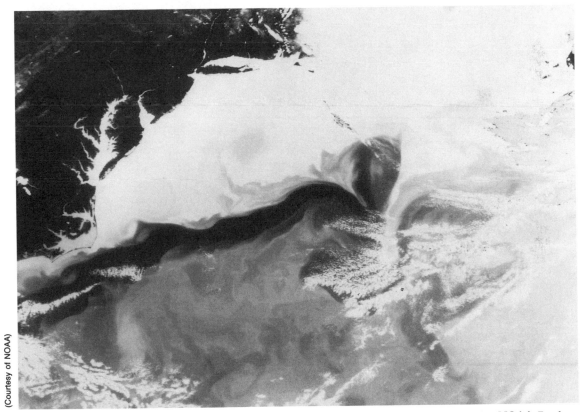

(Courtesy of NOAA)

FIG. 8-4.(b) Infrared image of the Gulf Stream off the east coast of the United States from the NOAA-7 polar-orbiting weather satellite. Commercial fishermen use such images to locate productive fishing grounds.

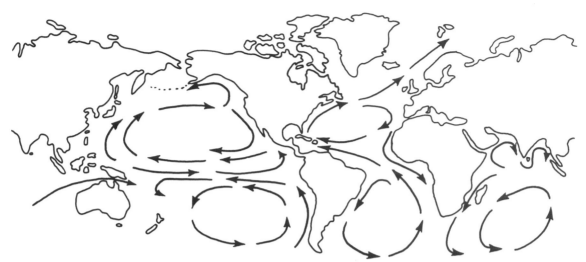

FIG. 8-5. The major ocean currents.

(Courtesy of NASA)

FIG. 8-6. A remarkable photograph of Hurricane Gladys, 150 miles southwest of Tampa, Florida, taken from *Apollo 7* on October 18, 1968.

anticyclonic, in the Northern Hemisphere, like that of a high-pressure weather system, and counterclockwise in the Southern Hemisphere.

The eddies are widespread, and have been identified in the midocean regions from the Arctic to Antarctica. They originate in far-reaching corners of the ocean and travel for great distances over long periods of time, up to several years. The eddies tend to trap marine organisms and transport them to hostile environments where they can only survive for as long as the eddies do. The eddies are apparently pinched off sections of a major ocean current and drift slowly along with the current. They also might form by strong ocean convection that is deflected by the Coriolis effect, by strong winds blowing across the surface, or by an abyssal current brushing against the bottom. Detecting eddies from satellites is possible because of the difference in temperature, density, and chemical and biological composition of eddies from that of the surrounding water. These differences might manifest themselves by a change in surface roughness, by sea surface temperature differences, or by differences in color as a result of the phytoplankton content of the sea water.

Surface winds are perhaps the single most important measurement of the ocean that can be made remotely from space. Backscatter from small wavelets and other surface features that are correlated with the wind are revealed on radar imagery. Each point on the ocean is measured from two different directions, to determine the wind speed and direction, which are in general agreement with ship and buoy wind reports. Unlike ships, which follow heavily used shipping lanes and therefore sample only a limited portion of the ocean, radar devices on satellites can cover the entire ocean every few days.

The surface winds are responsible for generating surface currents, which can be detected by radar altimetry. When a surface current is flowing, the earth's rotation causes the ocean surface to tilt slightly at right angles to the direction of flow. For major currents like the Gulf Stream, the difference in topography on the ocean surface can be as much as three feet. To detect these subtle changes in relief, the altimeter must be accurate to within a few inches though measured from several hundred miles in space from an orbiting satellite (FIG. 8-7).

The orbit of the satellite also must be determined accurately, as must areas where the Earth departs from a perfect sphere, or *geoid*, which is caused by gravitational and centrifugal forces. With further improvements, radar altimetry is capable of providing valuable information on global currents, which are important for distributing the ocean's heat around the world and which have a major effect on the climate.

MAPPING THE SEAFLOOR

The topography of the ocean surface measured by radar altimetry from GEOS-3 and Seasat satellites shows large bulges and depressions, with a relief as great as 600 feet. Because these surface variations are spread out over a wide expanse, however, they are undetected by the human eye. The shape of the surface is dictated by the pull of gravity from undersea mountains, ridges, troughs, and other structures of varying mass that are distributed unevenly over the ocean floor. The more massive the undersea mountain, the greater is the gravitational force, which makes the water pull in or pile up around the mountain, causing a swell in the ocean surface. Because submarine trenches have less mass to attract the water, troughs form in the sea surface over these undersea structures.

The satellite altimetry data is used to produce a map of the entire ocean surface (FIG. 8-8) that represents the ocean bottom as much as seven miles deep. Chains of midocean ridges and deep ocean trenches are delineated clearer than any other method of remote sensing the ocean floor. The maps uncovered new features such as rifts, ridges, seamounts (undersea volcanoes), and fracture zones, and they better defined features already known to exist. The maps also support the theory of *plate tectonics*, which holds that the Earth is broken into seven major plates that shift about, crashing into or moving away from one another as a result of seafloor spreading.

At midocean ridges such as the mid-Atlantic ridge, which runs down the center of the Atlantic Ocean and matches the curvature of the opposing coastlines of North and South America with those of Europe and Africa, upwelling magma generated in the Earth's interior erupts from the crest of the ridge, piling up volcanic rocks while pressing the plates on either side farther apart. As a result, the Atlantic is widening at the expense of the Pacific. On the western edge of the Pacific and elsewhere, deep undersea trenches consume the Pacific plate, which dives into the Earth's interior where it melts to supply new magma for volcanoes and spreading ridges.

The force of gravity pulling on the plate as it descends into the mantle also assists in the movement of the plates, which is generally less than a couple of inches a year. When two plates converge, mountains such as the Himalayas are forced up and continue to rise as long as the plates continue to push against each other. When two plates slide past each other, severe earthquakes can result. For example, the North American and the Pacific plates slide past each other creating the San Andreas fault system of southern California. When earthquakes occur on the ocean floor, they can produce destructive seismic sea waves, or *tsunamis*, that break on nearby shores.

POLAR ICE

The polar ice cover constitutes roughly 10 percent of the earth's surface and creates a major heat sink that helps drive the large-scale atmospheric and oceanic circulation (FIG. 8-9). Before the age of satellites, little was known about the seasonal variations in ice cover, and mariners

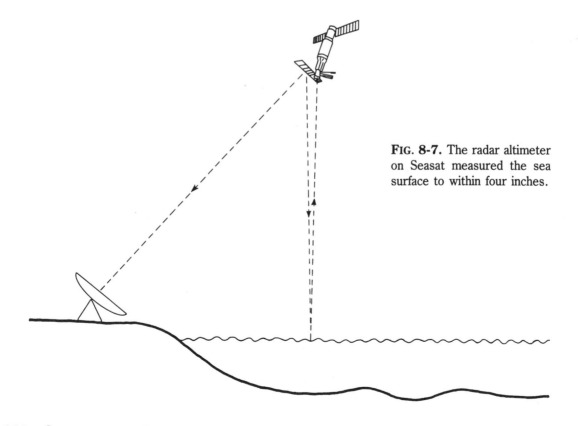

FIG. 8-7. The radar altimeter on Seasat measured the sea surface to within four inches.

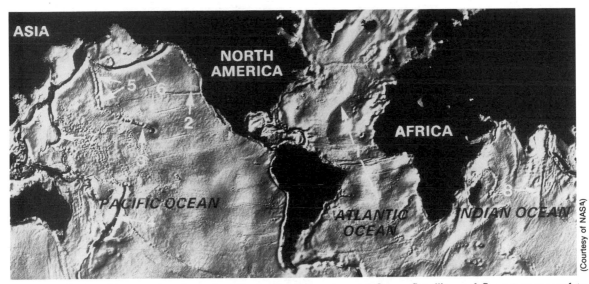

FIG. 8-8. Radar altimeter data from the Geodynamic Experimental Ocean Satellite and Seasat was used to produce this map of the ocean floor. 1-Mid-Atlantic ridge, 2-Mendocino fracture zone, 3-Hawaiian Island chains, 4-Tonga trench, 5-Emperor Seamounts, 6- Aleutian trench, 7-Marianas trench, 8-Ninety East Ridge.

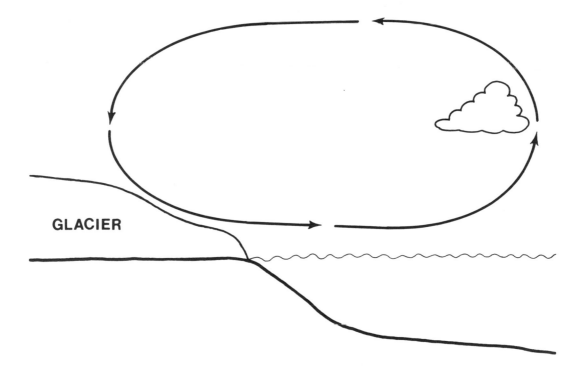

FIG. 8-9. The polar ice cover helps drive the large-scale atmospheric and oceanic circulation.

FIG. 8-10. NOAA-2 very high resolution radiometer image of pack ice along the Labrador coast.

took their chances against extreme cold, harsh winds, and constantly moving ice floes.

Observations from polar-orbiting satellites in the visible and infrared regions are severely hampered by darkness and frequent cloud cover, and consequently, this type of imagery is primarily used for large-scale observations of ice movement and extent (FIG. 8-10). Radar sensors on satellites provide all-weather, day and night observations of the polar ice. Radar image-tone variations and geometric shapes and forms help identify floes and ridges in the floating ice, as well as frozen leads and open water channels. Radar altimetry data which are accurate within an inch or so, also can delineate the boundary between arctic pack ice several feet thick and the surrounding sea. In addition, satellite radar observations can track the movement of ice floes, which can travel on average about ten miles a day. Because icebergs are an extreme hazard to shipping in arctic waters, satellite-borne radar can provide advance warning so that ship captains can find safer routes.

On the frozen continent of Antarctica, ice sheets averaging over one mile thick cover the entire land surface and constitute almost 90 percent of all ice on Earth. Antarctica annually discharges over 1 trillion tons of ice into the seas surrounding the continent, and pack ice covers an area of over 7 million square miles during the winter (FIG. 8-11). Seasonal variations in the ice cover are observed on visible and microwave imagery from polar-orbiting weather satellites.

The ice on West Antarctica's ice shelves is highly unstable, and a greenhouse warming of the climate brought on by increased carbon dioxide

FIG. 8-11. The extent of drift ice in Antarctica.

content in the atmosphere could bring the ice crashing down into the sea in a gigantic glacial surge. The increased sea level could set free more ice that would wander into the lower latitudes and melt. The rise in sea level by as much as 20 feet or more would inundate coastal regions, and many of the world's great cities, which are built by the sea or on inland waterways, would be flooded. Prime agricultural land located in low-lying areas would be under water, causing a shortfall in the food supply for many parts of the world. The sea ice also would have a major impact on the global climate because of an increase in albedo from sunlight reflecting off the bright ice. This possibility underscores the need for constant surveillance of the ocean by satellites, which might some day provide new insights into the relationship between sea and sky.

9

Looking at the Land

THE first aerial photographs of the land surface were taken from balloons with crude box cameras called *daguerreotypes*, named for the nineteenth century French inventor Louis Daguerre. The daguerreotype produced a photograph on a copper plate that was light-sensitized with a coating of silver iodide. These early cameras were often no more than a light - tight box with a pinhole or a simple glass plate for the lens. After a picture was taken, the photographic plate was removed from the camera, exposed to mercury fumes, and heated to produce a direct-positive image. Photography was further refined when the English inventor William Fox-Talbot developed the negative-positive process still in use today.

In the early 1850s, aerial photographs were used mainly for mapping and military reconnaissance. One of the first aerial photographs taken in the United States was of Boston Harbor in 1860.

Aerial photography and photographic interpretation for the military improved significantly during World War I with the use of aircraft. As cameras, films, and airplanes evolved, aerial photography was employed to satisfy an ever growing variety of needs. In the 1920s, vertical photographs made from aircraft became the basis of *photogrammetry*, the science of obtaining reliable measurements of objects on the ground from photographic images. Photogrammetry was further refined during World War II, for aerial reconnaissance of bomber targets (FIG. 9-1). Today, geologic, topographic, hydrologic, ecologic, and other types of maps are based solely on images of the Earth's surface made from aircraft and orbiting satellites.

TOPOGRAPHY

Topographic maps are a means of viewing the Earth's surface in three dimensions by drawing landforms using contour lines of equal elevations

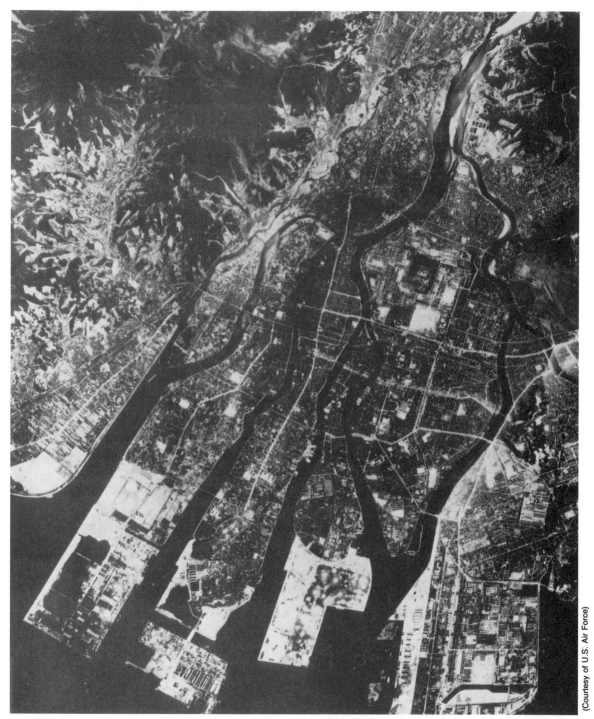

FIG. 9-1. An aerial photograph of Hiroshima, Japan, before the atomic bomb was dropped on August 6, 1945.

(Courtesy of U.S. Air Force)

(FIG. 9-2). They show more detail of natural and man-made features than do ordinary two-dimensional maps by depicting relative positions along with elevations. Topographic maps are useful to geologists and civil engineers, as well as many other users of maps who require that elevations are shown.

Prior to the turn of this century, topographic maps were made largely from observations on the ground. Today, such maps are made entirely by photogrammetric methods, and the only groundwork required is to field-check for accuracy. The maps are drawn using two aerial photographs taken at different points along the flight line, thereby

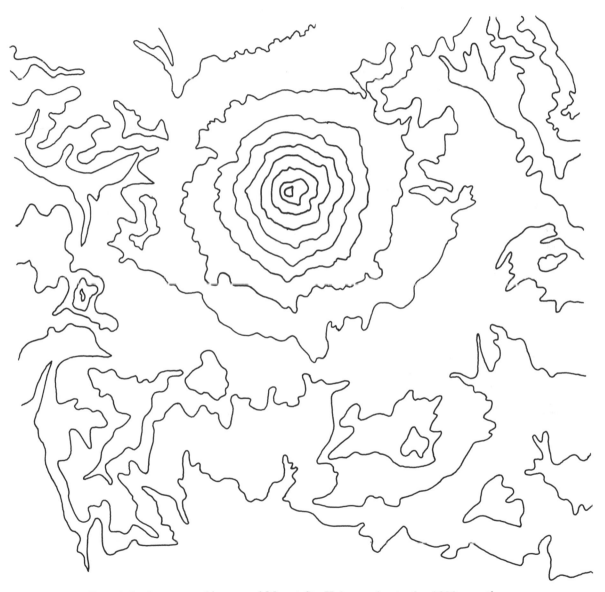

FIG. 9-2. A topographic map of Mount St. Helens prior to the 1980 eruption.

viewing objects on the ground at two different angles. The photographs then comprise a stereo pair that can be viewed with a stereoscope, which functions similar to a home stereoviewer that produces dazzling three-dimensional images. Topographic maps are made from a stereo pair by selecting points of equal elevation and feeding the data into a computer, which produces the map on a plotter. The maps are plotted with an accuracy that requires 90 percent of all well-defined features to be within $\frac{1}{50}$-inch of their true location on the map and the elevations of 90 percent of the features to be correct within 20 feet for 1:24,000-scale maps and within 50 feet for 1:62,500-scale maps.

One of the greatest limitations of space photography is its inability to view the ground in three dimensions. Such a capability would be of great value to inventorying, monitoring, and managing the Earth's land resources. One of the best clues toward the identification of a resource feature is its relief, which is not easily discerned in ordinary Landsat imagery. The problem is solved by making a synthetic stereo pair, with one image made in the conventional manner from Landsat multispectral scanner data and the other image made from the same data, except that stereoscopic parallax is artificially introduced into the scene so that features are perceived in three dimensions when the imagery is viewed through a stereoscope.

Stereophotography can also be made from satellite radar imagery viewed from different angles. Because radar is highly sensitive to changes in elevation, relief can be determined with a high degree of accuracy. Therefore, a great potential exists for using spaceborne radar data for three-dimensional analyses of the surface structure of Earth, as well as of other planets.

HYDROLOGY

Satellites can provide data on snow cover (FIG. 9-3), sea ice extent, river flow, and flood inundation. Regional snow cover is important for predicting the amount of runoff during the spring thaw. The Mississippi River (FIG. 9-4) is often beset with floods. The 1973 Mississippi River flood, the 1978 Kentucky River flood, and the 1978 Red River of the North floods resulted in near record-breaking water levels that were identified on satellite imagery. The unusually stormy winter of 1982–83, possibly as a result of El Niño, produced a huge mountain snowpack in the Colorado Rockies, and during the spring thaw, the Colorado River (FIG. 9-5) reached record flood stage.

Images obtained from GOES weather satellites are used to monitor river basin snow cover throughout the western United States and Canada. River basin snow maps are used by several governmental agencies, as well as private concerns such as utility companies. The snow data also are used to aid in dam and reservoir operations and to calibrate runoff models. These models are designed to simulate and forecast daily streamflow in basins where snowmelt is a major contributor to runoff. This is particularly important in the dry Western states where the data are used for preparing seasonal water supply forecasts.

Several methods are used to analyze GOES satellite data for mapping snow cover. The simplest uses an optical transfer device to magnify and rectify the satellite imagery so that it overlies a standard hydrologic basin map. The snow line is then transferred by hand from the image to the map. Another method uses a computer to display the image data on a video monitor, and the snow line is electronically traced onto the image. A third method uses a computer to determine the snow cover pixel by pixel by analyzing the terrain type and solar incidence angles.

The snow maps are prepared by regional offices of the National Environmental Satellite Data and Information Service (NESDIS) operated by NOAA. NESDIS prepares weekly snow and ice cover charts for the Northern Hemisphere during the snow season. The charts show the areal extent and brightness of continental snow cover but do not indicate the snow depth. The snow-cover charts are digitized and stored on computer tapes, and from these tapes, monthly anomaly, frequency, and climatological snow cover maps are created. In addition, continental or regional snow cover can be calculated over a long period of time for North American winter snow cover.

Satellite data are used routinely to detect and locate ice cover and ice dams on North American rivers. Observation of river ice is important because of the problems it creates for hydro-electric dams, bridges, and ship navigation. The ice becomes particularly hazardous when it breaks up and forms a dam, posing a flood threat to nearby communities (FIG. 9-6). Often, the ice persists because of river dams, sharp bends in the

river course, or branching of the main channel by islands. Landsat and GOES imagery can be used to detect and monitor ice jams, as well as monitor changes in river ice (TABLE 9-1).

Because GOES satellites are in geostationary orbits 22,300 miles above the equator, the imagery is limited to latitudes between 50 degrees north and south and to rivers over a mile wide. This precludes monitoring many subarctic rivers where ice is particularly troublesome. Fortunately, polar orbiting satellites such as NOAA and Landsat are useful for monitoring these areas. The satellites are also used for monitoring flash floods from large storm systems. Satellite-derived precipitation estimates and trends aid meteo-rologists and hydrologists in evaluating heavy

(Courtesy of NASA)

FIG. 9-3. A Skylab photograph of Mount Rainier in the Cascade Range of Washington showing snow cover.

precipitation events and providing timely warnings to the affected areas.

Flood damage in the United States often exceeds $1 billion annually, despite the construction of flood-preventing structures to help save lives and reduce losses. In order to minimize flood-related hazards, engineers and governmental officials need accurate information on the location of flood-hazard areas and assessments of areas of inundation when floods occur. Satellite imagery is a prime source for this information because of its continual coverage of the affected areas. In many cases, the flooded areas show up best in nighttime thermal infrared imagery because of the high temperature contrasts between land and water. Because water is such a good spectral

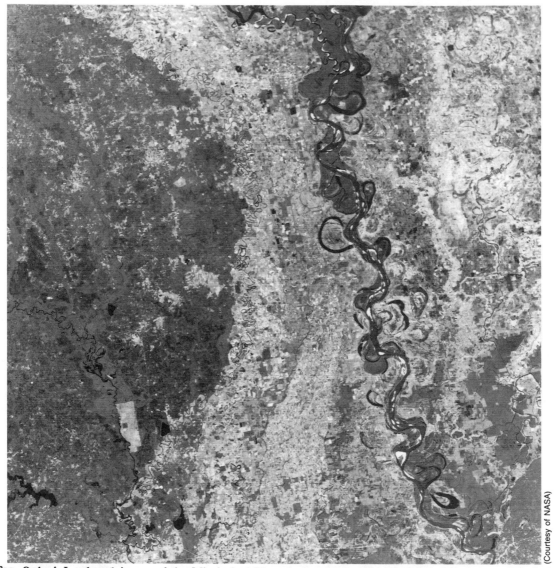

(Courtesy of NASA)

FIG. 9-4. A Landsat 1 image of the Mississippi River in Southeastern Arkansas, northwestern Mississippi, and northeastern Louisiana.

reflector of microwave energy, the flooded areas are easily delineated on satellite radar imagery both day and night and during cloud cover.

The NOAA advanced very high resolution radiometer (AVHRR) infrared data also can be used to identify areas inundated by large floods on large rivers. Computer models can be used to provide quick approximations of the total extent of a flood for disaster and relief planning.

ECOLOGY

Remote sensing of environmental changes on a global scale requires sensors that gather data over large areas often enough to provide continual

(Courtesy of NASA)

FIG. 9-5. A Skylab photograph of the Grand Canyon, as the Colorado River cuts its way through the earth's crust.

FIG. 9-6.(a) An ice jam in the Passumpsic River, causing flooding at St. Johnsbury Center, Vermont.

FIG. 9-6.(b) Very high resolution radiometer image of the Great Lakes and St. Lawrence Seaway showing widespread snow cover and frozen and unfrozen lakes and streams.

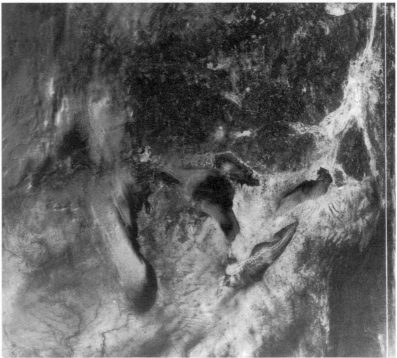

TABLE 9-1. Chronology of Major U.S. Floods.

DATE	RIVERS OR BASINS	DAMAGE IN $ MILLIONS	DEATH TOLL
1903	Kansas, Missouri and Mississippi	$ 40	100
1913	Ohio	150	470
1913	Texas	10	180
1921	Arkansas River	25	120
1921	Texas	20	220
1927	Mississippi River	280	300
1935	Republican and Kansas	20	110
1936	Northeast U.S.	270	110
1937	Ohio and Mississippi	420	140
1938	New England	40	600
1943	Ohio, Mississippi and Arkansas	170	60
1948	Columbia	100	75
1951	Kansas and Missouri	900	60
1952	Red River	200	10
1955	Northeast U.S.	700	200
1955	Pacific Coast	150	60
1957	Central U.S.	100	20
1964	Pacific Coast	400	40
1965	Mississippi, Missouri and Red Rivers	180	20
1965	South Platte	400	20
1968	New Jersey	160	----
1969	California	400	20
1969	Midwest	150	----
1969	James	120	150
1971	New Jersey and Pennsylvania	140	----
1972	Black Hills, S. Dakota	160	240
1972	Eastern U.S.	4,000	100
1973	Mississippi	1,150	30
1975	Red River	270	----
1975	New York and Pennsylvania	300	10
1976	Big Thompson Canyon	----	140
1977	Kentucky	400	20
1977	Johnstown, Pennsylvania	200	75
1978	Los Angeles	100	20
1978	Pearl River	1,000	15
1979	Texas	1,250	----
1980	Arizona and California	500	40
1980	Cowlitz, Washington	2,000	----
1982	Southern California	500	----
1982	Utah	300	----
1983	Southeast U.S.	600	20

coverage. Because of their near-polar, Sun-synchronous orbits approximately 500 miles above the Earth's surface, NOAA and Landsat satellites can provide this coverage. Satellite data have contributed significantly to inventorying and monitoring large ecosystems and land cover. The resolution and cost of the data enable investigators to create continental sized maps (FIG. 9-7) and global-sized maps both effectively and economically.

Different types of vegetation have different spectral signatures. Trees such as evergreens or conifers, appear dark red in multispectral color composites and remain that way throughout the year, whereas deciduous trees change from pink, to red, to brown during the seasons. Various colors are assigned to large-scale maps to distinguish major ecological units. For example, dark blue-green could be used for forests, dark blue for forested wetlands, dark red for unforested wetlands, and yellow for agricultural lands.

After each ecological unit is identified, its area is estimated to provide an areal inventory of land surface cover or ecosystem types. The data has become a highly useful tool for studies of arid environments and range-management programs.

Coastal wetlands form a highly productive buffer zone between the ocean and the upland. They also play an important role in estuary ecology and maintaining the quality of coastal environments. The need for a rapid, cost-effective method of mapping large tracks of wetlands necessitates the use of Landsat imagery.

The data is most useful when combined with other information, such as meteorological data and field observations. The information then can help resource managers update land use and resource inventories and correct boundaries between ma-

FIG. 9-7. Landsat mosaic of the United States made from 569 landsat images.
(Courtesy of The General Electric Company and NASA, copyright owned by The General Electric Company)

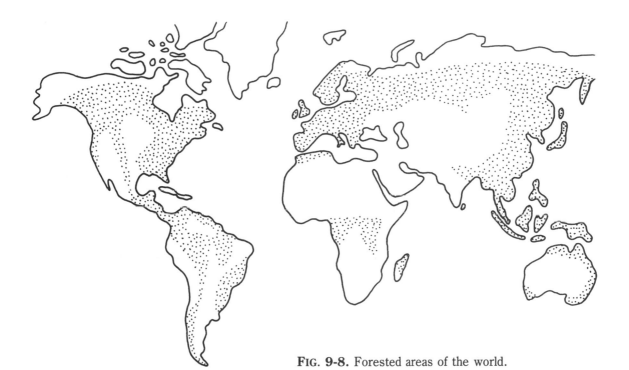

FIG. 9-8. Forested areas of the world.

jor ecological units. It also can facilitate mapping soil types and delineate physiographic provinces.

Once a database is established, monitoring land use and detecting change can be efficient and inexpensive. This information can then be stored in geographic information systems for comparison with later images to detect and measure rates of change or provide valuable current and historical information for land use planners and resource managers.

DEFORESTATION

Over the past two decades, the growth rate and general health of forests in the northeastern United States, eastern Canada, and many parts of central Europe have been declining. There are a number of factors working simultaneously to destroy the forests and streams and lakes, the most damaging of which is acid rain. The acids generated by the combustion of coal and petroleum run off into streams and lakes and percolate into the soil where they damage plant roots and leach out valuable soil nutrients. The direct contact of acids on foliage also destroys trees, as well as agricultural crops.

Tropical rain forests are also being destroyed, at the rate of about 30,000 square miles every year, by small-scale slash-and-burn agriculture and large-scale timber harvesting. If present trends of global deforestation continue, the accessible rain forests will be gone before the middle of the next century. Along with the forests will go a major portion of terrestrial species, which could mean mass extinction and a corresponding reduction in the Earth's genetic diversity. The world's forests (FIG. 9-8) are also a sizable reservoir of carbon, and the release of this carbon into the atmosphere by deforestation might have a major impact on the climate.

The development of a means of detecting, identifying, and quantifying symptoms of forest decline using satellite remote-sensing techniques is crucial. Such a system would provide investigators with the opportunity to assess and monitor forest destruction on a global scale.

Multispectral sensors on satellites such as Landsat and SPOT offer a perspective from which to study the Earth's vegetation that cannot be provided by any other means. A single Landsat image, which covers an area of 115 miles on a side, can take in an entire forest, and that same forest can be viewed every 16 days, which is the time it takes Landsat 5 to completely scan the Earth. Thus, Landsat provides a continuous coverage so that changing forest conditions can be detected. In addition, deforestation rates can be determined by comparing satellite imagery made at different times.

Vegetation stress generally reveals itself as changes in leaf structure, chlorophyll content, and water content, which produce spectral signatures that can be detected in computer-enhanced satellite multispectral imagery. The technique for measuring deforestation is basically quite simple. Healthy vegetation mostly absorbs red light and reflect infrared light, while bare ground mostly reflects red light and absorbs infrared light. When a patch of forest is cut down, the change in Landsat imagery is clearly visible, and by comparing two images pixel by pixel, the change can be measured with considerable accuracy.

In addition to mapping and monitoring forest decline and destruction, future satellite systems might allow investigators to detect spectral fingerprints associated with specific causes of deforestation such as disease, infestation, drought, or man.

URBAN DEVELOPMENT

As forested and agricultural lands give way to continued urbanization, the change in spectral characteristics is readily discernible on multispectral imagery from satellites. Up-to-date imagery of land use from Landsat is becoming increasingly necessary for planners to come to grips with the ever-growing problems of overpopulation on a planet with finite resources. The gathering of accurate data, which satellites can help provide, along with intelligent planning will ensure the best possible use of these valuable resources.

In urban areas, satellite imagery (FIG. 9-9) can identify different levels of habitation such as the central business district with a high density of buildings, dense residential areas with grass cover, and sparse residential areas with a moderate growth of trees. In many large American cities, a central core is surrounded by a prosperous and growing suburban and exurban region in a so-called *doughnut complex*. Built-up areas generally appear blue-gray on Landsat multispectral imagery because of the spectral signature of concrete. Grass yards and stands of trees appear red because of their high reflectance of near-infrared light. The comparison of imagery taken at various times can monitor urban sprawl as urban areas continue to expand.

For third-world countries, which have the largest birth rates as well as the highest death rates, remote sensing has become crucial for economic growth. Too often, economic strides made by poor countries are quickly eroded by the increase in mouths to feed. Remote sensing from aircraft and satellites can provide valuable information about available natural resources, which must be properly understood and managed if developing countries are to become self-sustaining.

The technology has been used by several developing countries, which have focused their developmental strategies on specific applications such as those directly related to energy sources. Most countries have come to realize that energy development is fundamental to the development of other natural resources, and this represents an asset that must be properly managed and sustained. Therefore, a means for rapid and comprehensive mapping of this important asset had to be found, and remote sensing has become that means for nearly every nation on Earth.

POLLUTION

As industrialization spreads throughout the underdeveloped countries of the world, so does pollution, which is an unfortunate by-product of prosperity. The richest nations of the world are also plagued with polluted air and water; contaminated soil and groundwater; and serious sewage, garbage, and toxic-waste disposal problems. One solution has been to pipe raw sewage directly into the ocean, dump garbage offshore, and burn toxic wastes on incineration ships out at sea. As a result of these actions, beaches in many parts of the world are unsightly and unsafe for swimming. Oil spills from oil tanker collisions and attacks by warring nations, along with offshore oil-well blowouts, seriously pollute nearby coastal areas and destroy the near-shore ecology.

Wastes in the ocean tend to concentrate in thermal layers and ocean fronts, which are also aquatic feeding grounds. Currents of the Gulf Stream laden with fish actually sweep over the dump sites, many of which are situated about 100

(Courtesy of NASA)

FIG. 9-9. Landsat 4 Thermatic Mapper image of the San Francisco Bay area.

miles off the eastern seaboard. Toxins, supposedly diluted to safe levels, are concentrated by biological activity, starting at the bottom of the food chain and working up to include man. Agricultural chemicals responsible for the green revolution are washing off the land into rivers, which empty into the oceans. As a consequence, seas like the Kattegat between Sweden and Denmark have lost much of their marine life. Tens of thousands of once healthy lakes in North American and Scandinavia are now totally devoid of fish as a result of the effects of acid precipitation.

Surface pollutants such as raw sewage, industrial effluents, and oil slicks can be detected by a variety of remote sensors on satellites. Oil slicks tend to have a higher index of refraction than background water at ultraviolet and visible wavelengths. In the thermal infrared region, oil has different radiation characteristics than water. Oil slicks dampen small waves, reducing the amount of radar backscatter, which is readily detectable on radar imagery. Because of their continual surveillance, satellites can track the drift and dispersion of oil slicks for cleanup efforts. Stream sediment load and suspended particulates are more reflective, and consequently, sediment-laden rivers and estuaries show up lighter on satellite multispectral imagery. Generally, dissolved substances absorb certain spectral bands and thus change the color of the backscattered light.

Not all pollutants are detectable by satellite, but they can be deduced by their association with other materials. Ocean dumping of wastes such as sewage and industrial acids are observed by Landsat and airborne sensors. Investigators can differentiate the types of wastes dumped into the ocean and determine their drift and dispersion. When used in conjunction with measurements taken on the surface, satellite sensors can enhance the ability to monitor certain pollutants in the water.

Air pollution around large cities and heavily industrialized areas is detectable on satellite imagery either directly or indirectly by its effects on the atmosphere. Air pollution consists of particulate matter from the combustion of fossil fuels; gases such as carbon monoxide, carbon dioxide, and ozone; and aerosols, which are suspended chemicals in the form of a fine dust or mist. In addition to man-made pollutants, there are natural pollutants from volcanic eruptions, forest fires, ocean salt spray, and dust storms. Organic compounds in the air interact with moisture and sunlight to form a photochemical smog that is prevalent in many urban areas.

Remote sensors can measure the concentration and movement of air pollution near urban areas, the chemical emissions from industrial plants, and other chemicals in trace amounts in the atmosphere. Atmospheric constituents such as water vapor, ozone, and carbon monoxide have been successfully measured from space. The degree to which thermal radiation emitted at the Earth's surface is attenuated during its passage through the atmosphere is directly related to the concentrations of these gases. Also, any increase in atmospheric carbon dioxide can lead to an increase in the opacity of the atmosphere and thus to an alteration in the spectral distribution of outgoing energy. An increase in atmospheric carbon dioxide from the combustion of fossil fuels and the destruction of forests and wetlands can have a dramatic effect on the climate, causing a change in precipitation patterns, increased desertification, and droughts in many parts of the world.

10

Mapping Minerals

THE insatiable requirement for fossil fuels and minerals to maintain a high standard of living in industrialized nations and to improve the standard of living in developing countries could lead to the depletion of known petroleum and high-grade ore reserves sometime during the first half of the next century. This potential depletion requires the search for new forms of energy and new ore deposits. Joining the geologist in this search for new energy and mineral deposits are several types of remote sensors on aircraft and satellites.

Mineral deposits reveal themselves in many different ways, most of which are invisible to the naked eye but are detectable in various sensors operating at wavelengths outside the visible spectrum. Satellite imagery can delineate geological structures such as faults, fracture zones, and contacts in which mineral ores are deposited and appear on satellite imagery as distinctive lineaments. Other structures such as folds or

domes (FIG. 10-1) are also distinguishable in satellite imagery and might serve as traps for oil and gas. Mineral deposits also might be detected as discolorations of the surrounding rock or by particular types of vegetative growth, which reflect certain soil types. The soils, in turn, are determined by their mineral content derived from the underlying parent rock.

Radar can penetrate heavy cloud cover and the vegetation canopy to observe the ground. Radar data is particularly useful for identifying structures and classifying rock units. Precision radar altimetry from satellites and other remote-sensing techniques can map the ocean bottom, where there is a large potential for the world's future supply of minerals and energy.

GEOLOGICAL STRUCTURES

Landsat imagery is being applied to a wide variety of geological problems that are difficult to solve using conventional methods alone. Remote sensing in geology is primarily used to augment conventional methods of compiling and interpreting geologic maps of large regions. Regional geologic maps present rock composition, structure, and geologic age, which are essential for constructing the geologic history of an area. Aided by geophysical data used for defining subsurface structures (FIG. 10-2), these reconstructions are important because formations of rock units and geological structures influence the deposition of mineral ores and petroleum.

Geologic maps incorporate field observations and laboratory measurements, which are limited by rock exposures, accessibility, and manpower. With remote-sensing techniques, however, it is possible to obtain certain structural and lithologic information much more efficiently than can be achieved on the ground. In well-exposed areas, geologic maps can be made from Landsat imagery even when only limited field data is available because many of the major structural and lithologic units are well displayed on the imagery.

The regional coverage of Landsat with an image area of 115 miles on a side, a low Sun angle, and repetitive coverage over the same site every 16 days make Landsat imagery particularly useful for geologic mapping. The low angle of illumination of the Sun is achieved by having the satellite cross the equator around 10:00 A.M. local time with each pass of its Sun-synchronous orbit. A low Sun

(Photo by J.R. Balsley, courtesy of USGS)

FIG. 10-1. Sedimentary dome at Sinclair, Wyoming.

angle casts shadows from geological structures, thus highlighting topographic relief. Shadowing is also useful for stereophotography from aircraft or synthetic stereoimagery from satellites.

Lineaments, which are long, linear trends in the Earth's surface, are one of the most obvious and most useful features in the imagery. Lineaments represent zones of weakness in the Earth's crust, often as a result of faulting. Other features frequently observed in Landsat imagery include circular structures (FIG. 10-3) created by domes, folds, and intrusions of igneous bodies into the crust. Stream drainage patterns, which are influenced by topographic relief and rock type, give more clues about the type of geological structure. In addition, the color and texture of the structure carry information about the rock formations that comprise it. With this information, it is possible to generate large-scale geologic maps of inac-

cessible areas from satellite imagery. The imagery is especially useful for geologic mapping of arid regions, where there is relatively little cloud cover.

In tropical regions having persistent cloud cover and heavy vegetation, radar imagery (FIG. 10-4) is used to acquire information about the geological structure that multispectral sensors are unable to. The geological structure is revealed by subtle changes in the height of the rain forest canopy, which conforms to the underlying topography. Also, certain microwave wavelengths can penetrate the vegetation and image the ground directly. As in Landsat imagery, lineaments and texture along with the *dip and strike* of the strata, which is the degree and direction of formation slope, aid in geologic mapping. Even with subdued eroded topography, major geological structures can be discerned. This is particularly important

FIG. 10-2. Seismic surveys are used to define subsurface structures.

(Courtesy of USDA)

in desert regions where radar can penetrate several feet into the dry sands to image morphological features such as old buried river channels and bedrock structures.

Subtle linear patterns are used to identify certain types of lithologic features that are not recognizable by other means. For example, large ignimbrite sheets composed of solidified deposits of volcanic ash in the Altiplano region of the Andes Mountains of South America were dissected by sets of straight, parallel gullies whose regional extent was unknown until they were structurally mapped from radar imagery. In like manner, faults such as those in California were detected by Seasat Synthetic Aperture Radar (SAR) and Shuttle Imaging Radar (SIR).

The geologic interpretation of the radar image is based on geometric shapes and patterns, including lineaments, joints, fracture patterns, folds, domes, drainage patterns, and other ero-

(Courtesy of NASA)

FIG. 10-3. The Brandberg structure in Namibia near Cape Cross in Southwest Africa is a zone of weakness in the Earth's crust caused by upwelling magma.

sional features, as well as the spatial relationships between these features. These patterns are interpreted in much the same manner as on other types of imagery with the advantage that the angle and direction of radar illumination can be selected. This factor is important for bringing out better topographic relief. Therefore, radar sensors are most useful for studying patterns and features that are expressed in changes of slope or surface roughness, and this is why radar has been used mostly for geomorphologic (landforms) and structural mapping.

The tone and texture in the radar image are used to discriminate between rock types. Different rocks are distinguished by differences in surface roughness and *dielectric constant*, the conductivity of the rocks, which is controlled by water content. These features affect radar backscatter, which in turn is readily delineated on radar images. Thus, radar aids in mapping lithological units over broad areas, but cannot be used to identify the rocks themselves.

Radar imagery enhances expressions of geological structure and rock formations, and can be used in conjunction with multispectral imagery. The two imaging techniques complement each other, and together, greatly improve regional exploration for oil traps and mineral deposits.

MINERAL EXPLORATION

As improvements are made in remote-sensing technology, it will become increasingly valuable in providing new tools for mineral exploration. Remote sensing from space can provide data for geologic mapping of unexplored regions (FIG. 10-5), mapping of fracture zones that control mineral deposition, and detection of hydrothermal alteration of host rocks associated with ore deposits. The broad coverage of satellite images facilitates geological studies on a regional scale. Computer-image enhancement makes it possible to emphasize features of interest, such as lineaments and surface roughness.

Lineaments, faults, fractures, and contacts, all of which control mineralization, are usually expressed on the surface as sharp changes in the surface topography, morphology, or vegetation cover. Lineaments in Nevada were mapped from

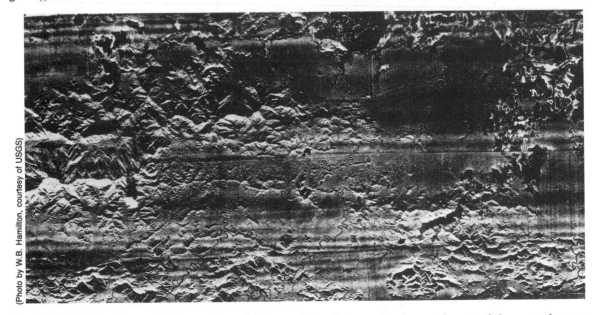

(Photo by W.B. Hamilton, courtesy of USGS)

FIG. 10-4. Side-looking radar image of central Papua, New Guinea, showing north part of the central range and south edge of the Sepik lowlands.

Landsat imagery and were shown to correlate well with the location of known mining districts. In central Colorado, a number of possible areas of interest were revealed, some of which were known to be mineralized while others remain to be explored.

Landsat false-color composites, in which individual spectral bands are ratioed and represented by a different color, can be used to locate areas of *hydrothermal alteration*, which is a discoloration of surface rocks by minerals such as limonite, a yellow-orange iron oxide mineral. Surface rocks are altered hydrothermally by hot mineralized water solutions often containing oxides and sulfides of iron that originate deep within the Earth's crust. The water is heated by a magma body and the

(Courtesy of NASA)

FIG. 10-5. A space shuttle photograph of the Himalaya Mountains in India and China.

heated water dissolves minerals in the crust. The mineralized fluids migrate toward the surface, deposit their mineral content, and return to the magma body to pick up more minerals.

Not all altered areas show up on the imagery, nor do all altered areas that do show up contain minerals in high enough concentrations to be considered ores. The detection of alteration depends on the reflectance spectra of the products of alteration. Thermal infrared scanners on aircraft can detect variations in the silica content of rocks, distinguish between different rock units, and identify areas of hydrothermal alteration, all of which are important factors in mineral exploration.

Combinations of multispectral and radar data can emphasize small differences in surface properties. The multispectral bands can discriminate geologic units on the basis of their reflectance, while radar data provide information about surface roughness. Composite images made from these data can discriminate subtle differences between surface deposits. These differences could go undetected by the geologist in the field.

The exploration of mineral deposits is significantly improved if lineaments and altered rock distributions from satellite imagery are combined with geophysical and geochemical data taken from the field. In areas where only reconnaissance geologic maps are available and accessibility is difficult, satellite imagery can be used initially to identify promising areas for more detailed mapping and geochemical surveys. These data, along with other geochemical and geophysical data, can identify specific areas that might have economic potential.

OIL AND GAS EXPLORATION

Landsat has made a major contribution to petroleum exploration, and many oil companies routinely use computer-enhanced satellite imagery to complement other exploration methods. Such methods include seismic surveys, which pass seismic waves, similar to sound waves, through the crust to image subterranean structures that might act as traps for oil. Seismic instruments are also towed behind ships (FIG. 10-6) to search for oil traps in the ocean crust on continental shelves.

Landsat imagery is used in a number of well-explored land regions, including the Rocky Mountains and Oklahoma (FIG. 10-7), in an at-

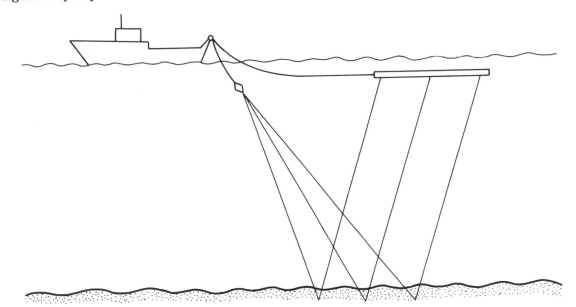

FIG. 10-6. A seismic survey of the oceanic crust.

tempt to better understand the accumulation of oil deposits in the area. The Landsat Thermatic Mapper (TM) imagery, with its larger number of spectral bands and higher resolution, can pick out subtle features that were not visible on earlier multispectral scanner (MSS) imagery. If an area looks interesting on MSS imagery, more detail can be brought out using TM imagery. Structural features such as folds, faults, dips, and strikes of particular rock formations, along with lineaments, landform features, drainage patterns, and other anomalies might suggest possible oil traps to the trained eye of the petroleum geologists. When all the evidence from various sources prove positive, a well is drilled, and with any luck, the sediments will contain oil.

Surface expressions such as domes, anticlines, synclines, and folds hold clues to the subsurface structure and are observed with spaceborne radar sensors because of the high sensitivity of the radar return signal to changes in slope. Tonal differences also can result from slope differences, which arise from the erosion of dipping beds. Various types of drainage patterns (FIG. 10-8) infer variations in the surface lithology. The drainage pattern density is anoth-

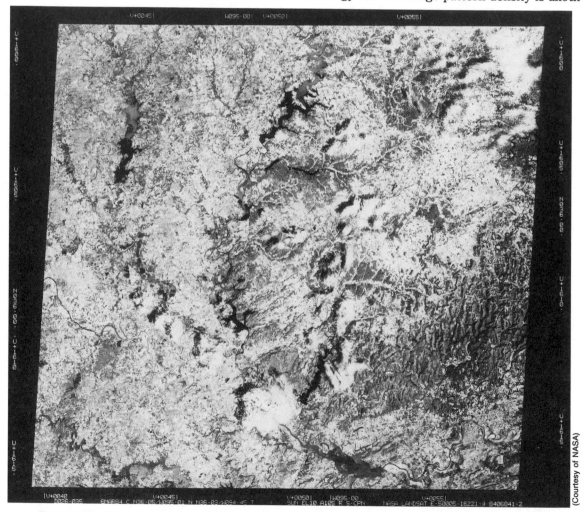

(Courtesy of NASA)

FIG. 10-7. A Landsat Thermatic Mapper image of the "oil patch" in the Tulsa, Oklahoma area.

er good indication of the lithology. Variations in the drainage density are also associated with variation in the coarseness of the alluvium.

In areas of exposed bedrock, drainage patterns depend on the lithologic character of the underlying rocks, the attitude of these rock bodies, and the arrangement and spacing of planes of weakness encountered by runoff. Any abrupt changes in the drainage patterns are particularly important because they signify the boundary be-tween two rock types. Drainage patterns are observed on radar imagery as a result of variations in local slope, variation in vegetation cover, and strong backscattering from boulders and pebbles in dry stream channels. In areas of extensive vegetation cover, drainage channels are observed mainly as a result of topographic expression.

Because cloud cover and heavy vegetation obstruct the view of the ground from Landsat, side-looking radar on aircraft and satellites can

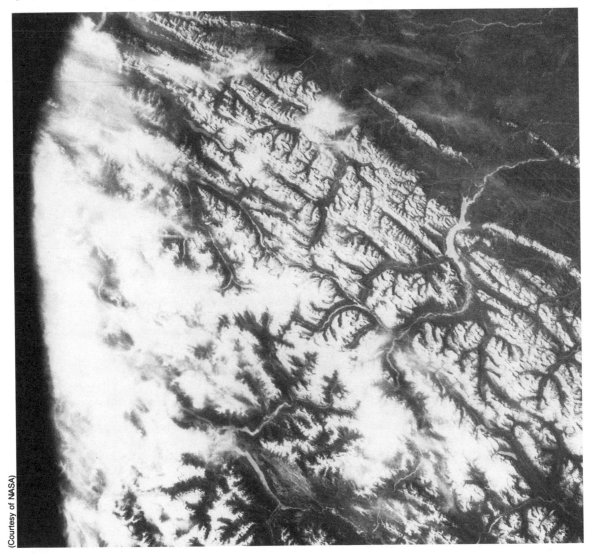

(Courtesy of NASA)

FIG. 10-8. Apollo spacecraft photograph of river drainage patterns in the Canadian Rockies in Alberta, Canada.

provide information about the underlying structure. This capability is particularly important in regions that cannot be explored by other methods. In the Amazon Basin of Brazil, mosaics made from airborne radar imagery formed base maps, which were employed as a guide for detailed mapping and field work in promising areas. The structural features provided information that greatly expanded the existing geological knowledge of the region. New geographical features were revealed, including previously unknown volcanic cones and uncharted rivers. In addition, large deposits of important minerals were discovered.

The angle of illumination of side-looking radar emphasizes landforms, and large areas that can be surveyed under constant conditions favor the recognition of extensive features that are too large-scale to be recognized by most other methods. With sophisticated computer-enhancing techniques, radar imagery can provide an essential

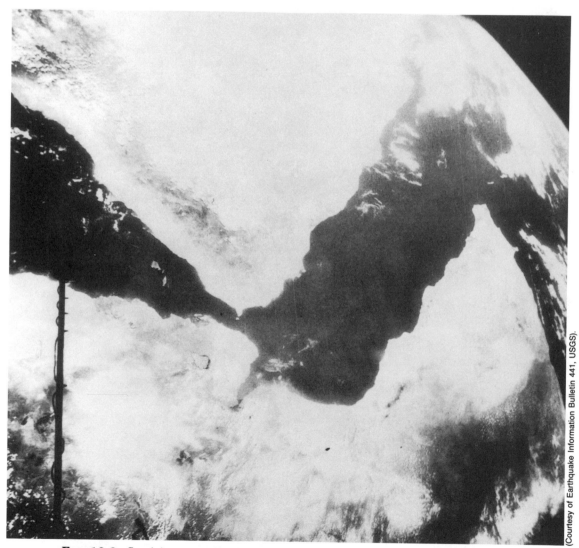

(Courtesy of Earthquake Information Bulletin 441, USGS).

FIG. 10-9. Gemini spacecraft photograph of the Red Sea and Gulf of Aden area.

exploratory tool, especially when its unique qualities are combined with multispectral satellite imagery. Together, these two imaging techniques can present a vivid representation of the land surface.

PLATE TECTONICS

The theory of plate tectonics offers many answers for why certain ore bodies exist where they are. Hydrothermal activity is a reflection of high heat flow, and high heat flow is associated with plate boundaries. The Salton Sea of Southern California and the Red Sea between Africa and Arabia (FIG. 10-9) show evidence of recently transported metals at extensional plate boundaries where the plates are pulling apart. Another area of high heat flow is associated with igneous activity at converging plate boundaries, where one plate is forced under another in subduction zones. Many hydrothermal ore deposits have been found in both new and ancient converging margins.

Extensive mountain building, volcanism, and granitic intrusion provide vein deposits of metallic ores. Some of the best gold deposits in the world are found in greenstone belts in northwest Canada that were invaded by hot magmatic solutions from granitic intrusions. *Ophiolite complexes*, which contain much of the world's ore reserves, are fragments of ancient undersea spreading centers shoved up on land when an oceanic plate dove under a continental plate.

Marine organisms buried under thick deposits of sediments were "cooked" into oil and gas millions of years ago by the high temperatures and pressures found in the Earth's interior. Continental drift and erosion of the land surface brought the oil and gas deposits to where they are today.

Presently, the Atlantic basin is widening and is spreading Europe and North America apart at a rate of about 1 inch per year, while the Pacific basin is being squeezed smaller. Measuring the rate of continental drift requires an extreme accuracy over a distance of thousands of miles. Standard geodetic survey methods cannot provide this accuracy, but measurements using satellites can. In Satellite Lasar Ranging (SLR), distances are measured by comparing how long it takes for laser light pulses to leave the ground, bounce off a satellite, and return to ground stations. In the satellite-based Global Positioning System (GPS), plate positions can be measured with a precision of about 1 inch over a distance of 300 miles. In Very Long Baseline Interferometry (VLBI), radio signals from distant quasars (rapidly spinning collapsed stars) are monitored at different stations on Earth. The difference in arrival times for these signals determine the distance between the receiving stations.

The VLBI method is the most exact of these techniques, having an accuracy approaching several parts per billion—comparable to measuring the length of 100 football fields to within the width of a human hair. These measurements of plate motion are in good agreement with geological methods, which are based on the spacing of magnetic stripes on the ocean floor caused by seafloor spreading. And seafloor spreading is the driving force behind plate tectonics, which is responsible for most of the Earth's landforms from magestic mountain ranges to giant rift valleys.

11

Assessing Agriculture

EVERY society, whether it is a primitive tribe or a modern western culture, tries to provide itself with the basic requirements of life: adequate food and shelter and a healthy environment. Only when these essential needs are assured can attention be turned to comfort and convenience, which determine a society's quality of life. Unfortunately, for much of the world, that quality of life suffers as people are forced to spend most of their efforts on obtaining enough food just to stay alive, with little income left for anything else.

A major part of the problem is that global military expenditures are sapping the economies of both large and small countries, preventing them from providing the necessary resources to meet even the most basic of needs. On the average, African countries spend four times as much on armaments as they do for agriculture. For many countries, the true threat to security is not war, but rather ecological deterioration, which is accelerating at a furious rate and manifests itself in vanishing forests and wetlands, topsoil depletion and desertification, improper irrigation methods and overuse of groundwater, population pressures on limited food resources and other natural resources, and the effects all these interrelated problems have on political and economic stability. It makes no sense to apply stopgap measures in order to keep up with a burgeoning world population, while destroying the very land required to feed that population. Perhaps this is where modern technology with emphasis on satellite surveillance can make the greatest contribution by analyzing the problem on a global basis and supplying answers that might make this a more secure world for everyone.

CLEARING THE LAND

As developing nations attempt to raise their standard of living, one of their first steps is to clear forests and wetlands for agriculture. Over 20 million acres of rain forest—about the size of the state of South Carolina—are cleared annually for agriculture and timber harvest. Much of the land is cleared by slash-and-burn methods, in which trees are set afire and their ashes used to fertilize the thin, poor-quality soil. After a year or two of improper farming or grazing methods, the soil is worn out, and the farmers press farther into the woods. The abandoned farms are then subjected to severe soil erosion because there is no longer vegetative cover to protect against the effects of wind and rain.

Lumber companies, using modern timber harvesting equipment that can snip trees at the base with giant shears and wood chippers that can reduce a 100-foot tree in half a minute, mow down the forests with comparative ease. Some forests are cut down to provide fuel for electrical power generation, which seems to be a tragic waste of a precious resource. In improverished countries without other fuels, trees are cut down to provide firewood for cooking and providing warmth.

The rain forests, which comprise only 6 percent of the land surface, are home to more than 65 percent of all terrestrial species. Some plants have medicinal values that might someday cure diseases like cancer and AIDS, and it would be criminal if they were to become extinct. Once the forests are leveled, the ground is often laid bare to the elements, and the soil is severely eroded. When the soil is gone, the forest has no chance of recovery and for all practical purposes is lost.

Swamps, marshes, fens, and bogs are perhaps the richest of all ecosystems, producing upwards of eight times as much plant matter per acre as an average wheat field. A large variety of plants and animals make wetlands their home. Coastal wetlands (FIG. 11-1) support valuable fisheries, and in the United States, about two-thirds of shellfish rely on these areas for spawning and nursery grounds.

Wetlands act as a natural filter, removing sediments and some types of water pollution.

(Courtesy of National Park Service)

FIG. 11-1. Mosquite Lagoon, Cape Canaveral National Seashore.

They also protect coasts against storms and erosion. Yet, wetlands are being destroyed for agricultural land at an alarming rate. Almost 90 percent of recent wetland losses in the United States has been for agricultural purposes.

The urgent need to feed hungry people is a major reason for draining the wetlands of the Third World. Short-term food production is replacing the long-term economic and ecological benefits of preserving the wetlands, with a consequential loss of local fisheries and breeding grounds for marine species and wildlife. In many cases, the destruc-

tion of the wetlands is irreversible, and for this very reason it is necessary that they stay wet.

Since Landsat imagery began in 1972, over 16 years of environmental destruction can be analyzed on a global scale. The disappearance of rain forests and wetlands is readily discernible on Landsat imagery, as a result of the lack of certain spectral qualities provided by trees and other vegetation that are no longer in existence. By going back over the earlier imagery, the progress of deforestation and wetland destruction can be followed (TABLE 11-1).

TABLE 11-1. Summary of Soil Types.

CLIMATE	Temperate (humid) > 160 in. rainfall	Temperate (dry) < 160 in. rainfall	Tropical (heavy rainfall)	Arctic or desert
VEGETATION	Forest	Grass and brush	Grass and trees	Almost none, no humus development
TYPICAL AREA	Eastern U.S.	Western U.S.		
SOIL TYPE	Pedalfer	Pedocal	Laterite	
TOPSOIL	Sandy; light colored; acid	Enriched in calcite; white color	Enriched in iron and aluminum; brick red color	No real soil forms because no organic material. Chemical weathering very low.
SUBSOIL	Enriched in aluminum, iron, and clay; brown color	Enriched in calcite; white color	All other elements removed by leaching	
REMARKS	Extreme development in conifer forest abundant humus makes groundwater acid; soil light gray due to lack of iron	Caliche (name applied to accumulation of calcite)	Apparently bacteria destroy humus; no acid available to remove iron	

The blackened ground from more recent forest burns is also highly visible in Landsat imagery. Roads along which settlers have cleared plots of land show up on the imagery as distinctive diagonal lines. In addition, forests fires from slash-and-burn agriculture, as well as natural causes, can be detected at night with infrared sensors on weather satellites. Color infrared imagery can provide an excellent contrast between upland and wetland vegetation for mapping wetland plant communities and determining their biomass. The soil surface and flooded areas under a heavy mangrove canopy can be mapped using radar on orbiting satellites. Microwave energy can penetrate the foliage, and the brightness of the radar image can be used to determine soil moisture.

The survival of rain forests and wetlands depends on protection through good resource management, which requires accurate and up-to-date information on large tracks of land that can only be supplied by remote-sensing techniques.

TOPSOIL DEPLETION

Improper farming and grazing methods in all parts of the world, but particularly in the developing countries, have destroyed once productive land by causing soil erosion (FIG. 11-2). The soil is layered into three zones (FIG. 11-3). The A horizon, or topsoil, is the best developed and most productive, with large amounts of organic nutrients required for healthy plant growth. The soil thickness ranges from a few inches to several feet, with an average worldwide thickness of only seven inches. The B horizon, or subsoil, is coarse, infertile soil that will not sustain good plant growth. The effect of soil

(Photo by Carrol Tyler, courtesy of USDA Soil Conservation Service)

FIG. 11-2. Severe soil erosion on a farm near Viola, Idaho.

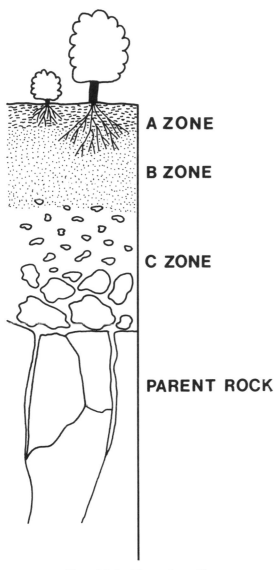

A ZONE

B ZONE

C ZONE

PARENT ROCK

FIG. 11-3. The soil profile.

erosion is to bring the B horizon to the surface, and because there are no plant roots to hold the subsoil, it also is eroded down to the parent rock.

The topsoil is disappearing at about three times faster than new soil is being generated by the breakdown of parent rock. It is estimated that 25 billion tons of topsoil is lost yearly. To put this number into perspective, imagine a mountain of dirt roughly 2 miles on a side and 2 miles high.

The soil particles are either carried away by the wind or washed off by the rain. They then enter rivers and streams and eventually make their way to the bottom of the ocean. The mud and silt also clog navigable river channels and build up behind dams, severely limiting their useful life.

In order to keep up with an ever-growing human population, which adds another 100 million mouths to feed every year, farmers have abandoned sound soil-conservation practices and have taken to more intensified farming methods. These methods include less rotation of crops, greater reliance on row crops, more plantings between fallow periods, and extensive use of chemical fertilizers instead of organic fertilizers (which hold the soil better). Almost all the good arable land is already under cultivation. Marginal lands, which are often hilly, dry, or contain only thin, fragile topsoils and therefore erode easily, are also forced into production.

During the decade of the 1970s, American farmers placed an additional 60 million acres—an area larger than the state of Kansas—into production in order to keep up with the increased world food demand. As a result, 3 billion tons of topsoil were eroded in 1977 alone. In the short term, these measures might generate substantial gains in productivity, but once the topsoil is reduced to only a few inches or is eroded entirely, crop productivity falls off dramatically and often irreversibly.

Another process that results from the loss of topsoil and takes millions of acres of once-productive cropland and pastureland out of production annually is called *desertification*. After the land is denuded of the clays and fine silts that constitute the topsoil, only the coarse sands are left behind. If this involves a large enough area, a man-made desert is created. The problem is exacerbated when the land is subjected to flash floods, higher erosion rates, and dust storms, which sweep the sands from one place to another. A classic example is the Sahel region of central Africa (FIG. 11-4), which once was heavily forested until nomads cut

FIG. 11-4. The Sahel region of central Africa.

down the trees to improve hunting and provide grasslands for their herds. As a result of overgrazing and drought conditions, the grasslands too disappeared and were overrun by the advancing sands of the Sahara Desert, which engulfed everything in their path and chased people out of the region.

The denuded land has a higher surface reflectance of sunlight, and this contributes to lesser amounts of rainfall and brings drought conditions, which denude more land and make the process of desertification self-perpetuating.

Satellites provide a powerful tool for monitoring deforestation, droughts, crop fluctuations, and other natural and man-induced processes worldwide. Another important agricultural application of satellite imagery is for mapping land-use potential and for monitoring water availability. In the developing world, crop statistics and soil mapping have proved to be difficult, labor-intensive, and time-consuming. Landsat imagery can delineate large plots of land, which then are cross-checked at selected sites in the field for accuracy. Accurate satellite image interpretation can detect agricultural growing conditions such as drought, disease, infestation, or frost damage.

Factors that affect plant vigor are reflected in the leaf structure, which in turn affects its spectral signature. As a plant matures, its color on a standard Landsat false-color composite

TABLE 11-2. Major Deserts of the World.

DESERT	LOCATION	TYPE	AREA SQUARE (Miles × 1000)
Sahara	North Africa	Tropical	3500
Australian	Western/interior	Tropical	1300
Arabian	Arabian Peninsula	Tropical	1000
Turkestan	S. Central U.S.S.R.	Continental	750
North America	S.W. U.S./N. Mexico	Continental	500
Patagonian	Argentina	Continental	260
Thar	India/Pakistan	Tropical	230
Kalahari	S.W. Africa	Littoral	220
Gobi	Mongolia/China	Continental	200
Takla Makan	Sinkiang, China	Continental	200
Iranian	Iran/Afganistan	Tropical	150
Atacama	Peru/Chile	Littoral	140

changes from pale pink to dark red. Unplanted fields appear blue or brown, depending on the type of soil. Each crop type has its own unique spectral signature, which can be used for broad-area identification (FIG. 11-5).

Once crops are identified, an assessment can be made for crop production. The crop production statistics are vitally important for developing countries because an unexpected shortfall accompanied by a delay in grain imports could result in famine.

DROUGHT

The recent droughts in Africa and the growing concerns over desertification have placed greater emphasis on the need for remote sensing from space. The sub-Saharan drought of the last quarter century has been the worst in 150 years, and the 1983 and 1984 droughts, which left upwards of a million people dead or dying from famine, were the worst in this century. One contributing factor to such a long-lasting drought could have been the stripping away of vegetation, which alters the reflective properties of the land. A greater influence might be an increase in atmospheric carbon dioxide content from the combustion of fossil fuels and the destruction of forests and wetlands, which released large quantities of stored carbon into atmosphere. In addition, the loss of vegetation means that less carbon dioxide is removed from the atmosphere by photosynthesis.

The accumulation of carbon dioxide in the atmosphere over the past century is expected to cause a general warming of the Earth through the *greenhouse effect* (FIG. 11-6) by trapping escaping radiant heat and reradiating it to the ground. The increase in global temperatures could affect the climate dramatically by shifting precipitation

FIG. 11-5. A Landsat image of San Joaquin Valley, California, showing cultivated fields enhanced for crop identification.

(Courtesy of NASA)

patterns throughout the world, which could bring unusually wet conditions to some areas and droughts to others.

One of the greatest developments in remote sensing is the ability to estimate the amount of green vegetation biomass and consequently terrestrial photosynthesis and primary production on a global scale. This ability could improve the understanding of the role of terrestrial plant life in the carbon cycle and allow monitoring of important changes in global vegetation and productivity that result from increasing atmospheric carbon dioxide and climatic changes.

Sensors on NOAA polar-orbiting weather satellites and the proposed Earth Radiation Budget Explorer (ERBE) satellite (FIG. 11-7) can collect daily high-resolution radiance data, which could be used to estimate the amount of solar radiation absorbed by vegetation. From this data, estimates of the green-leaf densities of plant canopies can be made to provide an accurate measurement of terrestrial photosynthesis. The satellite data then

can be correlated with observed atmospheric carbon dioxide fluctuations (FIG. 11-8) to determine whether fluctuations in the carbon cycle are a result of variations in photosynthesis.

The patterns show that terrestrial photosynthesis is inversely related to atmospheric carbon dioxide content. This relationship could greatly aid in understanding the greenhouse problem and detecting shifts in the distribution of plant communities caused by increasing carbon dioxide levels and by climatic changes.

Weather is the most important factor in the year-to-year changes in agricultural production, and accurate crop forecasts rely on routine weather observations from meteorological satellites. The satellite weather data are fed into numerical forecast models to provide climate predictions up to several months in advance. The weather data also are used in computer models to provide estimates of soil moisture, crop yield, and crop stress resulting from disease, infestation, or drought. In addition, infrared radiometers on polar-

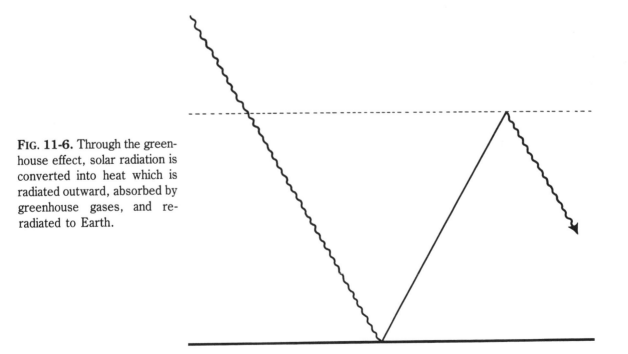

FIG. 11-6. Through the greenhouse effect, solar radiation is converted into heat which is radiated outward, absorbed by greenhouse gases, and re-radiated to Earth.

orbiting weather satellites can monitor the effects of drought on crops by classifying vegetation and estimating their amount of biomass per given area.

Meteorological factors that can be estimated from satellite data and are needed for crop models include precipitation, daily temperature extremes, canopy temperature (which is directly related to crop condition), and amount of *insolation*, or solar input. Because insolation is the primary energy source for growing crops, it is used to estimate crop yield, evaporation rates, and soil moisture, which can provide advance warning of damaging conditions. The direct measurement of the extent and condition of vegetation derived from satellite observations can, therefore, provide reliable warnings of impending drought and help identify those areas that will be hardest hit.

Deserts and the semiarid lands that surround them cover about one-third of the world's land surface. Because they are generally cloud-free, satellites have a clear view of desert floors. Satellite imagery is used to study drought-ridden areas in an effort to understand the process of desertification. These studies could help in-

FIG. 11-7. The Earth Radiation Budget Explorer satellite will anticipate climate trends that affect agriculture and other weather-related activities.

vestigators predict where the deserts of the future will be located and aid in finding ways of preventing soil erosion, loss of vegetation, and other desert-causing phenomena.

On enhanced color-infrared composites, land covered with vegetation appears red, while desertified areas appear white because of the high reflectance of sunlight. In areas that have suffered from prolonged drought, satellite imagery can detect changes in brightness, indicating that the soil is breaking down and the process of desertification is beginning. The spread of desertification can be traced through the migration of sand dunes (FIG. 11-9), which destroy everything in their paths.

By comparing present-day imagery with earlier imagery, investigators have found areas in the Sahel region of Africa that were once productive, but are now devoid of vegetation and topsoil. The satellite imagery also shows that rivers are much more turbid now than they were in the past because channels are forced to carry greater amounts of sediment from damaged areas. Some channels that formerly carried water year-round are now dry, which threatens river-bottom lands upon which millions of people depend for their survival.

IRRIGATION

Because Africa is prone to drought, it depends heavily on its water projects to irrigate the parched land. The Nile River Valley (FIG. 11-10) is one of the most developed irrigated areas in the world. Since the Aswan High Dam, which created Lake Nasser, was built in the mid 1970s, only a trickle of the Nile River has reached the Mediterranean Sea each year. Instead, most of its waters are used to irrigate some 20,000 square miles of farm land, encompassing an area about the size of West Virginia.

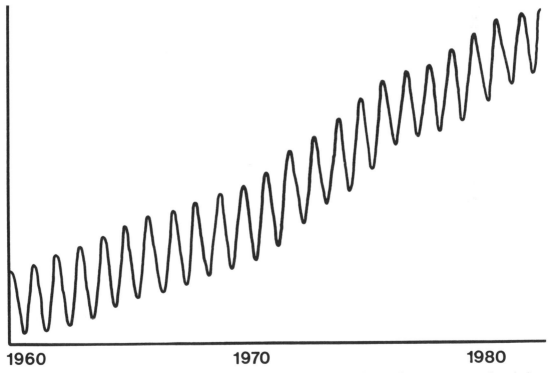

1960   **1970**   **1980**

FIG. 11-8. Carbon dioxide concentration in the atmosphere. The fluctuations are seasonal variations of carbon dioxide.

On the border between Zambia and Zimbabwe, the Zambezi River was dammed to create Kariba Lake, the largest artificial reservoir in the world. All the major rivers in South and Southeast Asia, including the Yangtze, the Mekong, the Irrawaddy, the Brahmaputra, the Ganges, and the Indus, also have been extensively developed for irrigation. China has the largest volume of regulated stream flow, which is used mostly for irrigation, than any other country in the world. Some 100,000 dams and reservoirs will give China a total storage capacity of about 100

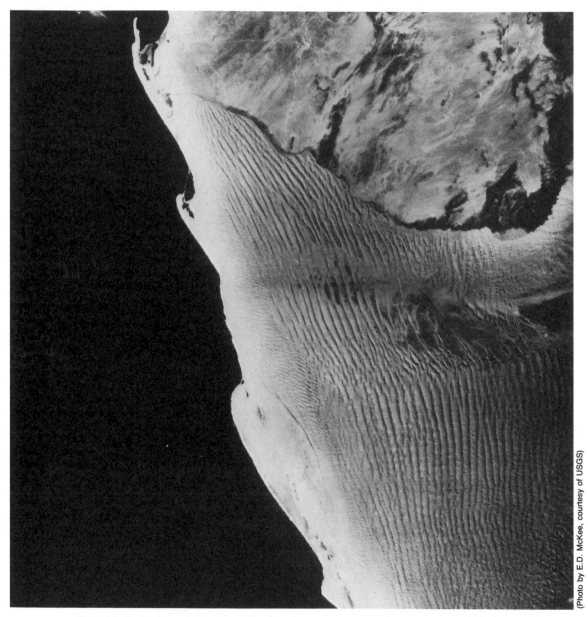

(Photo by E.D. McKee, courtesy of USGS)

FIG. 11-9. Landsat imagery of the northern part of the Namib Desert, Namibia.

cubic miles of water. The Soviet Union is planning to divert two of its largest rivers, the Ob and Yenisea, to the southwest across warmer, more arable land through a series of giant dams and canals.

Over 10 percent of the world's cultivated land is irrigated (FIG. 11-11), requiring some 600 cubic miles of water annually. The advantages of irrigation are that crops do not depend on the whims of nature for their water supply, more land can be brought under cultivation, and two or more crops can be grown in a single year. Its disadvantages are that most river water used for irrigation has a high salt content, and if fields are not drained properly, the salt buildup in the soil can ruin the land and cause crops to be stunted or die.

Tens of thousands of acres of once fertile land are destroyed by this process yearly, and estimates indicate that by the end of the century, over half of all irrigated land will be destroyed by salt (FIG. 11-12). With most of the good irrigation land

(Courtesy of NASA)

FIG. 11-10. Space Shuttle photograph of Egypt's Nile Valley.

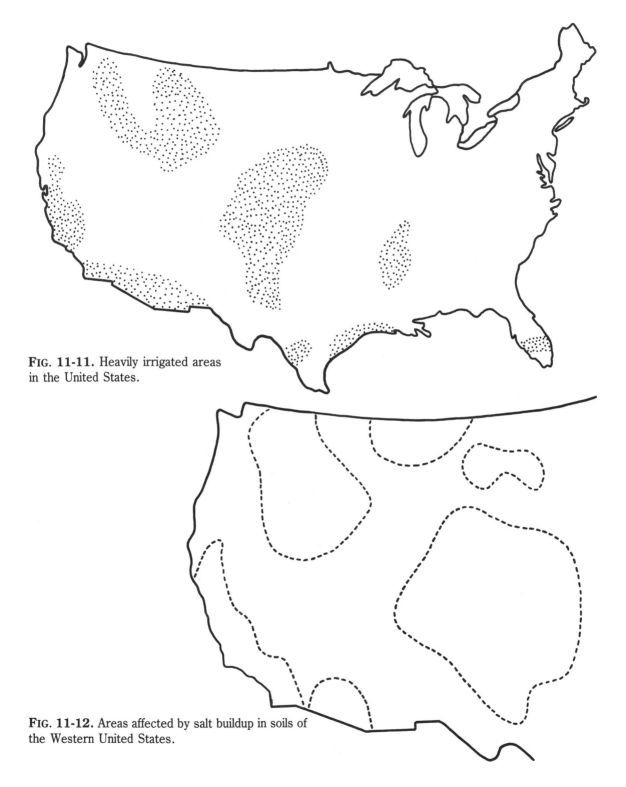

FIG. 11-11. Heavily irrigated areas in the United States.

FIG. 11-12. Areas affected by salt buildup in soils of the Western United States.

already in production, farmers can no longer afford to cultivate the land until it salts up and has to be abandoned. The problem is particularly bad in arid regions because there's only a small amount of rainfall, which is needed to flush out the soil. Overirrigation on poorly drained fields also can waterlog the soil and cause root rot, which also destroys crops. Agricultural chemicals such as fertilizers and pesticides are carried off with the drainwater, which ends up in rivers and streams and finally in the ocean where if the concentrations are high enough, the chemicals can kill fish and other marine life.

The second most important freshwater resource in the world is groundwater. An aquifer is composed of unconsolidated sand and gravel with connecting pore spaces. Gravity causes the water to flow through this formation at a rate of only a few inches per year. A water catchment at the head of the aquifer recharges the groundwater system. The rate of infiltration into the groundwater system depends on the distribution and the amount of precipitation, the type of rocks and soils, the slope of the land, and the amount and type of vegetation.

The use of groundwater for irrigation is expensive, and generally only affluent nations can afford to use it on a large scale. The water is pumped from wells, in some cases thousands of feet deep, by giant electric or diesel pumps, and is distributed over a circular area about half a mile across by a center pivot irrigation system (FIG. 11-13). In the great Kalansho Sand Sea of Libya, the water wells are drilled several miles apart (FIG. 11-14), and the desert is turned into pastureland for grazing sheep.

The overuse of groundwater can cause a lowering of the water table or total depletion of the aquifer. When an aquifer becomes depleted, subsidence causes compaction, which decreases the pore spaces between sediment grains. When this happens, the groundwater system cannot be restored to its original capacity and wells go dry.

PREDICTING FAMINE

The watch on world agriculture includes a systematic effort to use remote-sensing data to monitor crop development. There is too much at stake to leave the future to ordinary crop-forecasting methods. Too often food aid for Africa has been mobilized into action too late to forestall food emergencies with consequential loss of lives and dramatic increases in the expense of relief efforts. The 1984 disasters in Ethiopia and Sudan have spurred international organizations to try to improve systems to give early warning of food emergencies. In recent years, these agencies have turned to remote-sensing techniques to cut response time.

If predicting famine were a simple matter of detecting drought, satellite imagery would have solved the early warning problem. However, the causes of famine are more complex, requiring a knowledge of economic, political, and social factors, which vary from country to country. The Famine Early Warning System (FEWS) developed in 1986 by the U.S. Agency for International Development (USAID) is an attempt to create a program that takes the physical, as well as the human, factors into account. USAID's aim is to provide timely information that can be used by agency planners and government officials to meet food emergencies in African countries. The dependence on relief of these countries will continue to grow for as long as populations continue to grow and the land is laid waste by the activities of man and nature.

FIG. 11-13. A center-pivot groundwater irrigation system on a Kansas farm.

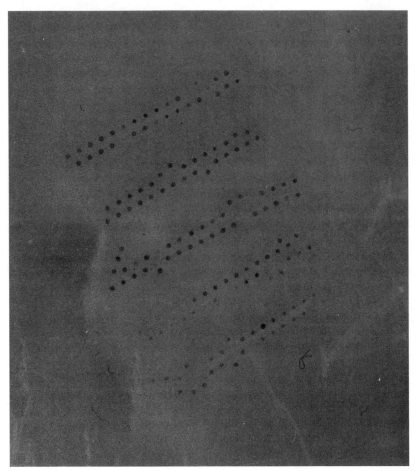

FIG. 11-14. A space shuttle photograph of circular irrigated areas in the great Kalansho Sand Sea of Libya.

(Courtesy of NASA)

12

Depicting Disasters

NATURAL disasters are uncontrollable occurrences that take many lives, destroy much property, and have been ongoing since the beginning of time. Before the age of technology, people had little or no warning of impending danger and therefore could not take defensive measures against the outbursts of nature. Because the Earth is dynamic and ever-growing, readjustments in the crust are often accompanied by earthquakes. If the earthquake epicenter happens to be near a city, buildings that have not been properly built or sufficiently strengthened will be destroyed, along with their occupants. Volcanoes make some of the most majestic mountains in the world. Yet, it is an ironic twist of nature that the most beautiful volcanic mountains also happen to be the most destructive.

Hardly a day goes by when there is not a major storm that takes lives and destroys property in some part of the world. Hurricanes and typhoons are among the most powerful forces in nature, producing the energy equivalent of a hydrogen bomb going off every minute they are alive. Floods caused by hurricanes and heavy thunderstorms destroy more property and take more lives than any other weather-related phenomenon. Landslides from storms and earthquakes are vivid reminders that the surface of the Earth is in constant motion. Although man has little control over nature and its destructive ways, he has the power to prevent the greatest catastrophe to himself and to the rest of the living world: the debacle of nuclear war.

EARTHQUAKES

Earthquakes are by far the strongest natural forces on Earth. In a matter of seconds, a large quake can level an entire city. Every year, a dozen or so major earthquakes strike somewhere in the world. Most of them are in areas along the rim of the Pacific plate, called the *circum-Pacific belt*. The Pacific Basin area is also prone to destruc-

TABLE 12-1. Earthquake Magnitude Scale and Expected Incidence.		
RICHTER SCALE	EARTHQUAKE EFFECTS	YEARLY AVERAGE
<2.0	Microearthquake-imperceptible	+600,000
2.0–2.9	Generally not felt but recorded	300,000
3.0–3.9	Felt by most people if nearby	50,000
4.0–4.9	Minor shock; damage slight and localized	6,000
5.0–5.9	Moderate shock; equivalent energy of atomic bomb	1,000
6.0–6.9	Large shock; possibly destructive in urban areas	120
7.0–7.9	Major earthquake—inflicts serious damage	14
8.0–8.9	Great earthquake; inflicts total destruction	1 every decade
9.0 and up	Largest earthquakes	1-2 every century

tive seismic sea waves from undersea earthquakes (TABLE 12-1).

The San Andreas fault system of southern California (FIG. 12-1) is also included in the circum-Pacific belt, and a major shock occurs there on average roughly once a year, thus making California among the shakiest ground on earth. California is also one of the most heavily populated regions in the United States and has a large proportion of the nation's high-tech industries. This puts space at a premium, and too often buildings, even hospitals, are built across faults.

Between now and the turn of the century, there is a better than fifty-fifty chance for a great earthquake, like the one that leveled San Francisco in 1906 (FIG. 12-2), to hit southern California. There is a near 100 percent probability that within the next 25 years, a major earthquake with a magnitude of 6 or better on the Richter scale will hit the eastern United States. Unless new structures are properly engineered and old structures sufficiently strengthened to withstand the effects of earthquake waves, the next earthquake to strike the United States could be monumental.

The San Andreas Fault is the best instrumented fault in the world. There are a number of remote-sensing techniques that are employed for making earthquake predictions. Laser ranging devices can measure the amount of crustal strain along the fault with an accuracy of ½ inch over a distance of about 20 miles.

There are several precursory signals that earthquake faults produce that might also aid in prediction. These signals include changes in the tilt of the ground, magnetic anomalies, increased radon gas content in nearby water wells, and swarms of microearthquakes. Faults also produce a phenomenon known as *earthquake lights* prior to and during rupture. Apparently, the strain on the rocks in the vicinity of the fault causes them to emit energy, which produces a faint atmospheric glow at night.

Seismic activity is associated with a variety of other electrical effects. A system of radio wave monitors distributed along the San Andreas Fault have recorded changes in atmospheric radio waves prior to earthquakes that occurred between 1983 and 1986. The increased electrical conductivity of rocks under stress near the fault apparently

FIG. 12-1.(a) Streams offset by the San Andreas Fault of Southern California.

FIG. 12-1.(b) The San Andreas Fault system.

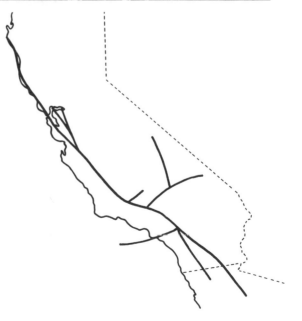

FIG. 12-2. Destruction from the 1906 San Francisco earthquake.

(Courtesy of NOAA)

causes radio waves to be absorbed by the ground one to six days prior to an earthquake. In addition, investigators observed short pulses of increased radio interference caused by the release of electromagnetic energy from cracking rocks.

Over the past few years, *geodesists* (scientists who measure the Earth) have significantly improved the precision with which they can determine positions on the Earth's surface using the satellite-based Global Positioning System (FIG. 12-3). With this system, investigators can monitor the strain accumulation on the San Andreas Fault with less cost than standard geodetic methods. Positions are determined with an accuracy of about 1 part in 10 million over a distance of up to 300 miles. The orbit of the satellite is precisely determined by comparing satellite signals received at two stations on the ground, and this in turn provides data on their relative positions.

There are presently six GPS satellites in orbit, and a full constellation of 18 satellites, scheduled to continue launching in 1989, will keep six satellites in sight of any point on the globe at all times for increased accuracy. In addition to monitoring the San Andreas Fault, the satellites

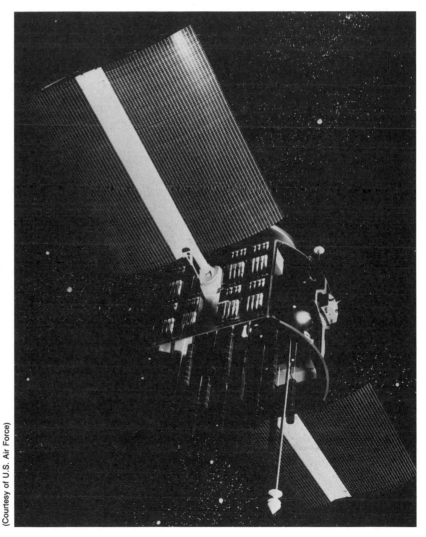

FIG. 12-3. Satellite of the Global Positioning System.

(Courtesy of U.S. Air Force)

can measure the rate of seafloor spreading in Iceland, the subsidence of the crust resulting from the removal of groundwater, and the bulging crust above the magma chamber under Long Valley California.

VOLCANOES

Like earthquakes, volcanoes are associated with crustal movements and occur on plate margins. When one crustal plate dives beneath another, the lighter rock component melts and rises to the surface in blobs called *diapirs*. Diapirs provide the magma for volcanoes and other igneous activities.

Another type of volcano, called a *hot spot volcano*, exists in the interior of a plate and arises from magma originating deep inside the mantle. Volcanic islands such as the Hawaiian Islands are formed conveyor-belt fashion as the oceanic crust travels over the hot spot.

Subduction zone volcanoes such as those in the western Pacific and in Indonesia are among the most explosive in the world, creating new islands and destroying old ones. The reason for their explosive nature is that the magma contains large amounts of volatiles consisting of water and gases. When the pressure is lifted as the magma reaches the surface, these volatiles are released explosively, fracturing the magma, which shoots out of the volcano like pellets from a shotgun. The Cascade Range of the western United States is composed of a chain of volcanoes extending from northern California to Canada. These volcanoes are associated with a subduction zone under the North American continent. The May 18, 1980, eruption of Mount St. Helens (FIG. 12-4), which devastated 200 square miles of national forest, is a good example of the explosive nature of these volcanoes (TABLE 12-2).

Many of the same methods used to predict earthquakes can be applied to predict volcanic eruptions. When a volcano is on the verge of erupting, there might be a swarm of micro-earthquake activity, which can be detected with seismometers. There might be moderate-size earthquakes associated with the eruption, although they are not nearly as great as those produced by faults. As the magma chamber swells with new magma, tiltmeters placed in strategic locations around the volcano can indicate whether it is on the verge of eruption. The Global Positioning System mentioned in the section on "Earthquakes" might also be able to detect slight bulges in the crust prior to a volcanic eruption. The flowing magma disturbs the magnetic field around the volcano, which can be detected by airborne magnetometers. The movement of magma causes slight changes in the gravity field that could be detected using gravimeters. There are also changes in electrical currents in the crust, which can be measured by resistivity meters. Other indications of impending eruption include a rise in temperature of near-surface rocks, hot springs, and gas vents. An analysis of volcanic gases for variances in trace elements such as gold, platinum, and iridium might indicate whether magma was moving into the volcanic system.

Indonesia is known for its explosive volcanoes, like Tambora and Krakatoa, two of the world's greatest eruptions. The 1982 eruption of Galunggung and the 1983 eruption of Una Una produced thick ash clouds that clogged jet airliner engines, causing them to stall. These situations prompted an investigation of the use of satellite observations for detecting and tracking volcanic clouds so that air traffic could be rerouted around them.

Possibly the dirtiest volcano of this century was the El Chichon eruption in southeastern Mexico on March 28, 1982. Almost immediately after each of the three major eruptions of this volcano, a large ash cloud appeared in GOES and NOAA weather satellite imagery. Over the next several weeks, the stratospheric dust cloud was observed to encircle the Earth in a narrow belt and then diffuse to the lower and upper latitudes until it completely covered the planet.

The dust cloud was more closely watched than that of any other volcano. A vast array of observations were taken from the ground, aircraft, high-altitude balloons, and satellites. There was ample reason for these observations: the effect the dust cloud had on the global climate. By absorbing and reflecting sunlight (FIG. 12-5), the dust cloud could have cooled the Northern Hemisphere by as much as ½ degree centigrade. This translated into a cool, damp summer, and an unusually cold winter in 1983. In addition, the eruption might have triggered a coincidental occurrence of the 1982–83 El Niño, which played havoc on the weather in many parts of the world.

VIOLENT STORMS

In late August 1900, a tropical storm began to brew over the mid-Atlantic Ocean and gathered considerable strength as it traveled westward. In early September, it raked several Caribbean islands on its way to the Gulf of Mexico (FIG. 12-6). With just two ship sightings to go on, the United States Weather Bureau issued severe storm warnings for the Gulf Coast area. The forecasters knew the hurricane was out there somewhere, but they had no idea of its size, location, or direction of travel. Then on the evening of September 8, with no warning, the hurricane charged into the Texas resort town of

(Courtesy of NASA)

FIG. 12-4. Devastation from the May 18, 1980, eruption of Mount St. Helens.

DATE	VOLCANO	AREA	DEATH TOLL
A.D. 79	Vesuvious	Pompeii, Italy	16,000
1669	Etna	Sicily	20,000
1815	Tambora	Sumbawa, Indonesia	12,000
1822	Galung Gung	Java, Indonesia	4,000
1883	Krakatoa	Java, Indonesia	36,000
1902	La Soufriere	St. Vincent, Martinique	15,000
1902	Pelee	Pierre, Martinique	28,000
1902	Santa Maria	Guatemala	6,000
1919	Keluit	Java, Indonesia	5,500
1985	Nevado del Ruiz	Armero, Columbia	20,000

TABLE 12-2. The Ten Most Wanted List of Volcanoes.

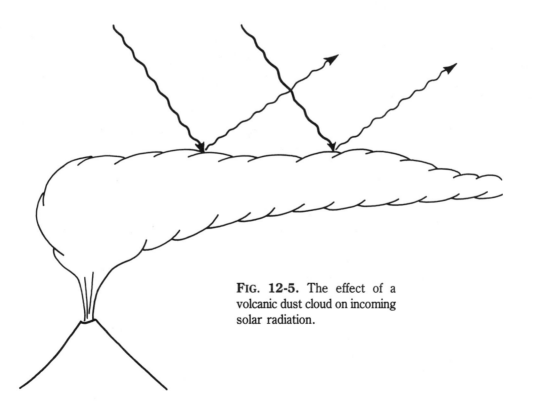

FIG. **12-5.** The effect of a volcanic dust cloud on incoming solar radiation.

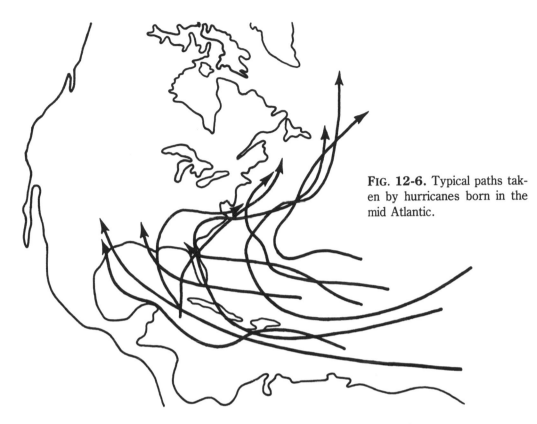

FIG. 12-6. Typical paths taken by hurricanes born in the mid Atlantic.

Galveston, submerged the only bridge connecting Galveston island with the mainland, and stranded its residents. The next morning, after the waters receded, the beaches were littered with thousands of human and animal corpses. The death toll was estimated at some 12,000 people, 6,000 in the city alone. It was the worst natural disaster in the nation's history.

Sixty one years later, another dangerous hurricane, named Carla, bore down on the Gulf Coast, only this time it was watched closely by weather satellite, which not only could locate the storm but could estimate its size, intensity, and movement. This prompted timely hurricane warnings and led at that time to the largest evacuation on American soil. The evacuation was highly successful with the loss of only 46 lives.

Bangladesh is one of the poorest nations on Earth, with a population of 100 million crammed into an area the size of Wisconsin, one-third of which is under water during the annual monsoon floods. Millions of farmers attempt to survive on low-lying islands and along coastal floodplains in the broad river delta of the Ganges and Brahmaputra rivers. On November 13, 1970, an immense cyclone, born in the Indian Ocean, headed up the Bay of Bengal (FIG. 12-7). Although the cyclone was tracked by weather satellite and storm warnings were issued, the government was powerless to take any action because there was no means for transporting large numbers of people to safety. In addition, peasants were reluctant to leave their homes for fear that squatters would take over their land. Winds clocking over 100 miles per hour pushed up 50-foot tidal waves that surged into the bay and swept over islands and coastal plains. With no high ground to take refuge to, people were forced to ride out the storm the best they could. Settlements were smashed, boats were overturned, vital fishing grounds were destroyed, and whole islands disappeared. When the storm finally passed, it left in its wake close to 1 million people dead. It was the worst storm-related catastrophe in modern history.

Violent storms are by far the costliest and deadliest natural disasters on Earth, destroying billions of dollars worth of property (FIG. 12-8) and taking thousands of lives in all parts of the world annually. One of the most important contributions made by satellites is detecting and tracking tropical cyclones, which include hurricanes in the Atlantic, Gulf Coast, and eastern Pacific (FIG. 12-9); typhoons in the western Pacific; and cyclones in the Indian Ocean. Since the beginning of the U.S. Operational Satellite Ser-vice in February 1966, no tropical storm has gone undetected anywhere in the world.

Weather satellites also provide near-continuous coverage of atmospheric conditions that produces tornadoes, thunderstorms, and other local severe convective storms. Telltale signs of strong convection often can be detected by satellites before severe storms develop. The developing storms then are monitored and tracked, and if necessary, the National Weather Service sends out severe storm warnings. American satellite data are also used extensively by foreign weather services, including those in the Southern Hemisphere, where relatively few conventional weather data are available and typhoons and monsoon floods are common.

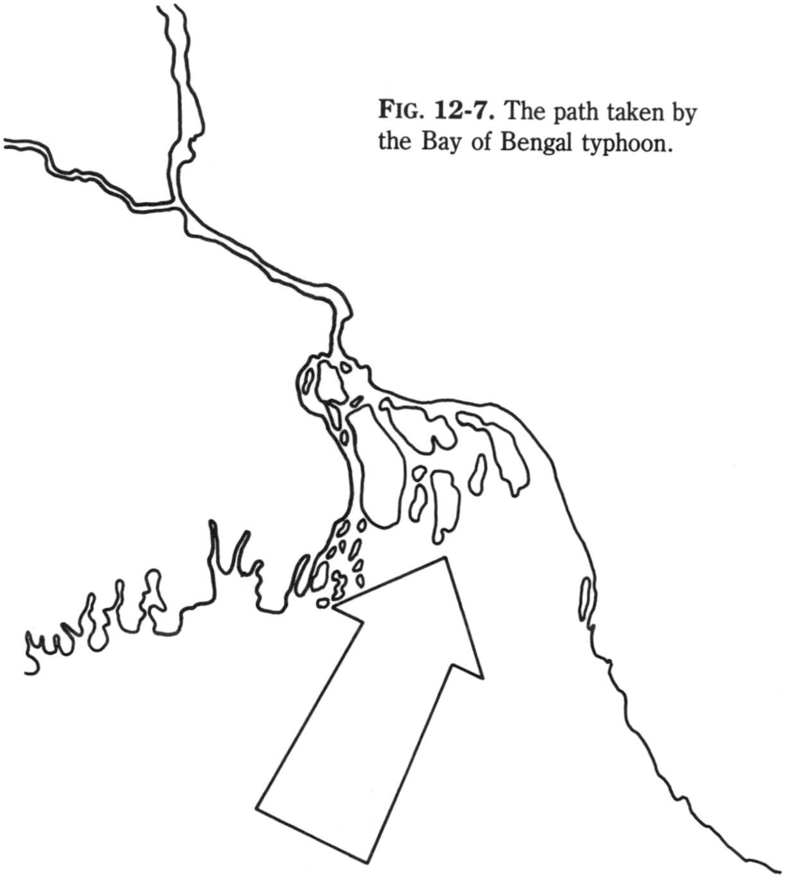

FIG. 12-7. The path taken by the Bay of Bengal typhoon.

FLOODS

Floods are natural recurring events that only become a disaster when people build on floodplains. The purpose of the floodplain is to carry excess water when a river overflows its channel during a flood. Levees are often built to contain a river at flood stage, placing the river higher than the adjacent floodplain. The worst flood-related catastrophe in modern times occurred in 1887 when the Yellow River overflowed its levees and flooded much of northern China, drowning 7 million people in what has been called "China's Sorrow."

Floodplains provide level ground and fertile soil for agriculture, but they also attract commerce. Rapid development of these areas without full consideration of the flood potential can end in a man-made disaster. Since 1936, the U.S. federal government has spent nearly $10 billion on flood-control projects. Although flood-pre-venting structures have helped to protect lives and reduce property losses, flood damage in the United States often exceeds $1 billion annually.

Satellites can provide improved information on the location of flood-hazard areas and make assessments of areas of inundation when floods occur. The 1973 Mississippi flood and the 1978 Pearl River flood in Louisiana and Mississippi were two of the costliest floods in American history, causing over $1 billion in damage each. In both cases, the flooded areas showed up best in satellite thermal infrared imagery taken at night because of the high temperature contrast between land and water.

Because water is such a good spectral reflector of microwave energy, satellite radar also can be a useful tool for determining the extent of flooding. With the use of computers, the satellite information can provide speedy estimates of the

(Courtesy of NOAA)

FIG. 12-8. Remains of a 32-unit apartment building in Gulfport, Mississippi from Hurricane Camille in August 1969.

total areal extent of a flood for disaster planning and relief efforts. Satellite imagery also can be used to determine the amount of winter snowpack, which in turn is used to estimate spring runoff and annual flooding from meltwater. Weather satellites can track thunderstorm development which might result in flash floods in the affected areas. In some cases, the deluge can break apart large chunks of land, causing destructive earth movements.

Heavy downpours are the primary cause of *ground failures*, which are violent shifts of the earth by water-logged soils. Ground failures also are produced by earthquakes and volcanic eruptions. In many parts of the world, these catastrophic events have resulted in major losses of life. One of the worst landslides on record resulted from the 1920 Kansu, China, earthquake, which induced several ground failures as much as 1 mile

FIG. 12-9. A GOES weather satellite image of Hurricane Allen in the Gulf of Mexico and Hurricane Isis off Baja California.

in length and breadth that killed as many as 180,000 people. When the Nevado del Ruiz volcano in Columbia erupted in 1985, it sent a 130-foot wall of mud crashing down the mountain side into the town of Armero, killing over 22,000 people. Although in the United States, most landslides take place in rural areas with no major loss of life, damages to property amount to over $1 billion annually.

Californians are all too aware of destructive landslides in their state. In the last decade, there have been thousands of landslides in the Los Angeles basin alone. In many cases, the landslides have sent homes careening down hillsides.

WARS

All the destruction wrought by nature throughout man's existence on Earth could not approach the potential devastation by today's stockpiles of nuclear and biochemical weapons. In this age when the United States and the Soviet Union can push a button and cause the deaths of over half the world's human population, not to mention the deaths of huge numbers of other species, our greatest protection lies in satellites. They can give early warning of enemy missile attacks, thus providing vital information for determining the type of retaliatory response. This ability thereby reinforces deterrence.

Spy satellites, which have dazzling resolution and can listen in on communications and missile telemetry, monitor the military activities on both sides. This makes satellites a powerful tool for verifying arms control agreements (FIG. 12-10) which in the past, did not work effectively because of mutual fear and distrust that one side would cheat and therefore gain advantage. Satellites also can keep an eye on the clandestine activities of other nations. For instance, on September 22, 1979, the surveillance satellite Vela observed what was reported as a "nuclear event" near Prince Edward Island in the Southern Ocean off South

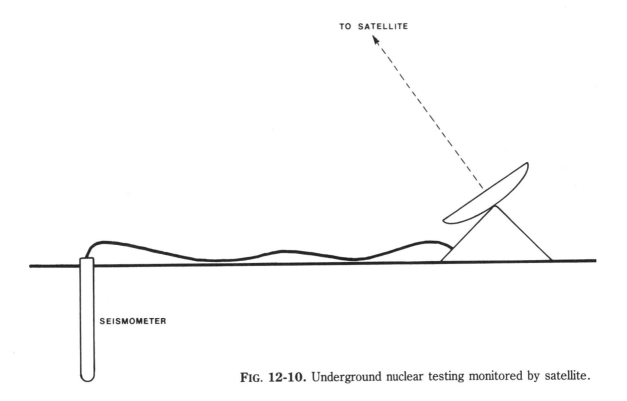

FIG. 12-10. Underground nuclear testing monitored by satellite.

Africa. This could justify fears that Third World countries, which have a greater propensity for going to war with each other, also might have nuclear weapons.

For every weapon there is a counterweapon, and this has spurred the development of weaponry down through the ages. The best counterweapon against nuclear missiles is thought to be the space-based nuclear defense system called the Strategic Defense Initiative (SDI), or more popularly known as "Star Wars." One idea is to place lasers, particle-beam accelerators, or kinetic launchers on a string of orbiting satellites that can engage enemy missiles during their boost phase, when they are the most vulnerable. The problem with this type of weapon system is that it can be highly destabilizing, creating more tension between the superpowers and bringing nuclear war

closer to reality. There is also a counterweapon against SDI. Killer satellites and ground-launched antisatellite missiles can be used to destroy the satellite defense system prior to hostilities.

The enormous cost for complete national protection could place a severe drain on the economy. As a result, a compromise system might be needed, in which only missile sites are given protection to safeguard their deterrence capability, still leaving the civilian population unprotected.

If deterrence fails and a full-scale nuclear war ensues, the effects to the Earth could be similar to the environmental destruction that caused the extinction of the dinosaurs and three-quarters of all other species 65 million years ago. During that time, the Earth might have been struck by a large meteor (FIG. 12-11) that lofted large quantities of dust and soot into the atmosphere and blocked

FIG. 12-11. Skylab photograph of the Manicouagan impact structure in Quebec Province, Canada. The reservoir was created when a huge meteor blasted out a crater some 60 miles across about 200 billion years ago.

out the Sun, just as would thousands of nuclear detonations.

The evidence for such devastation is contained in 65-million-year-old sediments throughout the world that consist of a thin layer of mud, formed when the dust cloud settled out. The mud had a high concentration of iridium, an isotope of platinum that is rare on Earth but relatively abundant in meteors. In addition, there is a thin layer of common soot, which might have been deposited from globalwide forest fires set ablaze by the meteor impact. Without life-giving sunlight, plants, which are a basic component in the food chain, would have died along with the animals that fed on them. The collision could have caused a worldwide downpour of acid rain, which would have killed plants along with aquatic life. The dust cloud also could have killed off most of the marine algae. The algae are the chief source of atmospheric dimethyl sulfide, which functions as a nucleus for cloud formation. Without clouds to reflect the Sun's rays back into space, the Earth would have heated up substantially, making conditions intolerable for most lifeforms.

With this spectacle in mind, modern technology can provide the greatest contribution toward safeguarding the Earth. Satellites are the best technical means for making a timely global assessment of man's activities in all parts of the world. These assessments do not appear to be very promising, however. Human populations are continuing to grow at a great expense to the rest of the world, inflicting a global impact of unprecedented dimensions.

The intricate interrelationships among species are not yet fully understood, but it is becoming more apparent that dramatic changes are taking place, and that our species is responsible for many of those changes. The loss of species diversity through environmental degradation could leave the Earth in a lower state of biological activity, which in turn could pose a great threat to humans. Nuclear war would be the ultimate insult to nature, bringing on an environmental catastrophe unequaled in the long history of Earth. But the insatiable human consumption of the earth, in the long run, can be equally catastrophic. Perhaps by using the most sophisticated technology ever devised by man for the purpose of exploring the Earth, we will learn from past mistakes and intelligently correct the most difficult problems that have ever faced our planet.

Bibliography

Selected references are provided for further reading.

DISCOVERY OF SPACE

Van Allen, James A. ''Myths and Realities of Space Flight.'' *Science* Vol. 232 (May 30, 1986): 1075-1076.

Burbidge, Margaret E. ''Adventure into Space.'' *Science* Vol. 221 (July 29, 1983): 421-426.

Logsdon, John M. ''U.S.-European Cooperation in Space Science: A 25-Year Perspective.'' *Science* Vol. 223 (January 6, 1984): 11-16.

Meredith, Dennis. ''How new 'eyes' track the mysterious origins of cosmic rays.'' *Popular Science* Vol. 228 (March 1986): 68-71.

Moore, Patrick. *Space: The Story of Man's Greatest Feat of Exploration*. Garden City, N.Y.: Natural History Press, 1969.

Scott, Arthur F. ''The Invention of the Balloon and the Birth of Modern Chemistry.'' *Scientific American* Vol. 250 (January 1984): 126-129.

Von Braun, Wernher, and Frederick I. Ordway III. *History of Rocketry & Space Travel*. New York: Thomas Y. Crowell Co., 1975.

PROBING THE PLANETS

Arvidson, Raymond E., Alan B. Binder, and Kenneth L. Jones. ''The Surface of Mars.'' *Scientific American* Vol. 238 (March 1978): 76-89.

Eberhart, Jonathan. ''Watering Mars with volcanism.'' *Science News* Vol. 132 (July 4, 1987): 9.

Gold, Michael. ''Voyager to the Seventh Planet.'' *Science 86* Vol. 7 (May 1986): 32-39.

Haberle, Robert M. ''The Climate of Mars.'' *Scientific American* Vol. 254 (May 1986): 54-62.

Ingersoll, Andrew P. ''Jupiter and Saturn.'' *Scientific American* Vol. 245 (December 1981): 90-108.

Johnson, Torrence V., Robert Hamilton Brown, and Laurence A. Soderblom. "The Moons of Uranus." *Scientific American* Vol. 256 (April 1987): 48-60.

Kerr, Richard A. "Venusian Geology Coming into Focus." *Science* Vol. 224 (May 18, 1984): 702-703.

Laeser, Richard P., William I. McLaughlin, and Donna M. Wolff. "Engineering Voyager 2's Encounter with Uranus." *Scientific American* Vol. 255 (November, 1986): 36-44.

Prinn, Ronald G. "The Volcanoes and Clouds of Venus." *Scientific American* Vol. 252 (March 1985): 46-53.

Soderblom, Laurence A., and Torrence V. Johnson. "The Moons of Saturn." *Scientific American* Vol. 246 (January 1982): 101-116.

Waldrop, M. Mitchell. "Voyage to a Blue Planet." *Science* Vol. 231 (February 28, 1986): 916-918.

Wilson, Lionel, and James W. Head III. "A comparison of volcanic eruption processes on Earth, Moon, Mars, Io, and Venus." *Nature* Vol. 302 (April 21, 1983): 663-669.

EYES IN THE SKY

Abelson, Philip H. "Spacelab." *Science* Vol. 225 (July 13, 1984): 3.

"Assessing the Earth." *Sky & Telescope* Vol. 71 (May 1986): 447.

Eberhart, Jonathan. "IRAS satellite to be 'revived' for tests." *Science News* Vol. 127 (January 26, 1985): 54.

Kerr, Richard A. "Monitoring Earth and Sun by Satellite." *Science* Vol. 236 (June 26, 1987): 1624-1625.

Neugebauer, G., et al. "Early Results from the Infrared Astronomical Satellite." *Science* Vol. 224 (April 6, 1984): 14-21.

Pollack, L., and H. Weiss. "Communications Satellites: Countdown for INTELSAT VI." *Science* Vol. 223 (February 10, 1984): 553-559.

Roller, Norman E. G., and John E. Colwell. "Coarse-resolution Satellite Data for Ecological Studies." *Bioscience* Vol. 36 (July/August 1986): 468-472.

Waldrop, M. Mitchell. "First Image from SPOT." *Science* Vol. 321 (March 28, 1986): 1504.

Yulsman, Tom, "Experiments in Space." *Science Digest* Vol. 92 (July 1984): 39-44.

DETECTION FROM A DISTANCE

Giacconi, Riccardo. "The Einstein X-Ray Observatory." *Scientific American* Vol. 242 (February 1980): 80-102.

Habing, Harm J., and Gerry Neugebauer. "The Infrared Sky." *Scientific American* Vol. 251 (November 1984): 49-57.

Henbest, Nigel. *Mysteries of the Universe* New York: Van Nostrand Reinhold, 1981.

Meredith, Dennis. "How new 'eyes' track the mysterious origins." *Popular Science* Vol. 228 (March 1986): 68-71.

Raloff, Janet. "Ozone Depletion's New Environmental Threat." *Science News* Vol. 130 (December 6, 1986): 362-363.

Roller, Norman E. G., and John E. Colwell. "Coarse-resolution Satellite Data for Ecological Studies." *Bioscience* Vol. 36 (July/August 1986): 468-472).

Scientific American, The Mind's Eye, New York: W. H. Freeman, 1986.

IMAGE INTERPRETATION

Braconnier, L. A., and P. J. Wiepking. *Introduction to the U.S. Geological Survey's EROS Data Center*, Washington, D.C.: Government Printing Office, 1978.

Cannon, T. M. "Image Processing by Computer." *Scientific American* Vol. 245 (October 1981): 214-225.

"Eosat Will Market Landsat Data From Chinese Ground Stations." *Aviation Week & Space Technology* Vol. 127 (July 20, 1987): 52.

Goetz, Alexander F. H., Greg Vane, Jerry E. Solomon, and Barrett N. Rock. "Imaging Spectrometry for Earth Remote Sensing." *Science* Vol. 228 (June 7, 1985): 1147-1153.

Hardisky, M. A., M. F. Gross, and V. Klemas. "Remote Sensing of Coastal Wetlands." *BioScience* Vol. 36 (July/August 1986): 453-458.

Maslowski, Andy. "Eyes on the Earth." *Astronomy* Vol. 14 (August 1986): 9-10.

Raloff, Janet. "IR can spy plant stress before eyes do." *Science* Vol. 127 (February 2, 1985): 70.

Waldrop, M. Mitchell. "An Inquiry into the State of the Earth." *Science* Vol. 226 (October 5, 1984): 33-35.

REMOTE SENSING BY RADAR

Adams, R. E. W., W. E. Brown, Jr., and T. Patrick Culbert. "Radar Mapping, Archeology, and Ancient Maya Land Use." *Science* Vol. 213 (September 25, 1981): 1457-1463.

Bower, Bruce. "Shuttle Radar is Key to Sahara's Secrets." *Science News* Vol. 125 (April 21, 1984): 224.

Elachi, Charles. "Radar Images of the Earth from Space." *Scientific American* Vol. 247 (December 1982): 53-61.

——————. "Spaceborne Imaging Radar: Geologic and Oceanographic Applications." *Science* Vol. 209 (September 5, 1980): 1073-1082.

Elachi, Charles, JoBea Cimino, and M. Settle. "Overview of the Shuttle Imaging Radar-B Preliminary Scientific Results." *Science* Vol. 232 (June 20, 1986): 1511-1516.

Jensen, Homer, L. C. Graham, Leonard J. Porcello, and Emmett N. Leith. "Side-looking Airborne Radar." *Scientific American* Vol. 237 (October 1977): 84-95.

Jones, W. Linwood, Peter G. Black, Victor E. Delnore, and Calvin T. Swift. "Airborne Microwave Remote-Sensing Measurements of Hurricane Allen." *Science* Vol. 214 (October 16, 1980): 274-280.

WATCHING THE WEATHER

Brownlee, Shannon. "Forecasting: How Exact Is It?" *Discover* (April 1985): 10-16.

Gleik, James. "They're Getting Better About Predicting the Weather." *New York Times Magazine* (January 27, 1985): 30-45.

Hallgren, Richard E. *Operations of the National Weather Service.* Washington, D.C.: U.S. Government Printing Office, 1985.

Hughes, Patrick. "Weather Satellites Come of Age." *Weatherwise* (April 1984): 69-75.

Kerr, Richard A. "Weather Satellites Coming of Age." *Science* Vol. 229 (July 19, 1985): 255-257.

Killinger, Dennis K., and Norman Menyuk. "Laser Remote Sensing of the Atmosphere." *Science* Vol. 235 (January 2, 1987): 37-44.

Schneider, Stephen H. "Climate Modeling." *Scientific American* Vol. 256 (May 1987): 72-80.

Smith, W. I., et al. "The Meteorological Satellite: Overview of 25 Years of Operations." *Science* Vol. 231 (January 31, 1986): 455-462.

OBSERVING THE OCEAN

Bartusiak, Marcia. "Mapping the sea floor from space." *Popular Science* Vol. 224 (February 1984): 81-85.

Bernardo, Stephanie. "The Seafloor: A Clear View from Space." *Science Digest* Vol. 92 (June 1984): 44-48.

Crane, Robert G. "Remote Sensing and Polar Climate." *Earth and Mineral Sciences* Vol. 55 (Spring 1986): 38-41.

Fiedler, Paul C. "Satellite Observations of the 1982-1983 El Niño Along the U.S. Pacific Coast." *Science* Vol. 224 (June 15, 1984): 1251-1254.

Fieldman, Gene, Dennis Clark, and David Halpern. "Satellite Color Observations of the Phytoplankton Distribution in the Eastern Equatorial Pacific During the 1982-1983 El Niño." *Science* Vol. 226 (November 30, 1984): 1069-1070.

Gannon, Robert. "Solving the puzzle of El Niño." *Popular Science* Vol. 229 (September 1986): 82-86.

Kerr, Richard A. "Small Eddies Are Mixing the Oceans." *Science* Vol. 230 (November 15, 1985): 793.

Perry, Mary Jane. "Assessing Marine Primary Production from Space." *BioScience* Vol. 36 (July/August 1986): 461-466.

Weisburd, Stefi. "Sea-surface shape by satellite." *Science News* Vol. 129 (January 18, 1986): 37.

Yates, H., et al. "Terrestrial Observations from NOAA Operational Satellites." *Science* Vol. 231 (January 31, 1986): 463-469.

LOOKING AT THE LAND

Clark, Carolyn, et al. "Mapping and classifying large ecological units." *BioScience* Vol. 36 (July/August 1986): 476-477.

Hardisky, M. A., M. F. Gross, and V. Klemas. "Remote Sensing of Coastal Wetlands." *BioScience* Vol. 36 (July/August 1986): 453-458.

Kerrod, Robin. *NASA Views of Earth*, New York: Gallery Books, 1985.

Kiehl, J. T. "Satellite Detection of Effects Due to Increased Atmospheric Carbon Dioxide." *Science* Vol. 222 (November 4, 1984): 504-506.

Peterson, Ivars. "Watch on Acid Rain: A Midterm Report." *Science News* Vol. 132 (January 18, 1987): 36.

Rock, B. N., et al. "Remote Detection of Forest Damage." *BioScience* Vol. 36 (July/August 1986): 439-444.

"Seeing the Forest." *Scientific American* Vol. 255 (September 1986): 66.

Sternlieb, George, and James W. Hughes. "The Changing Demography of the Central City." *Scientific American* Vol. 243 (August 1980): 48-53.

Taranik, James V., and Mark Settle. "Space Shuttle: A New Era in Terrestrial Remote Sensing." *Science* Vol. 214 (November 6, 1981): 619-626.

MAPPING MINERALS

Elachi, Charles. "Spaceborne Imaging Radar: Geologic and Oceanic Applications." *Science* Vol. 209 (September 5, 1980): 1073-1082.

Flinn, Edward A. "Application of Space Technology to Geodynamics." *Science* Vol. 213 (July 3, 1981): 89-95.

Goetz, Alexander F. H., and Lawrence C. Rowan. "Geologic Remote Sensing." *Science* Vol. 211 (February 20, 1981): 781-790.

Goetz, Alexander F. H., et al. "Imaging Spectrometry for Earth Remote Sensing." *Science* Vol. 228 (June 7, 1985): 1147-1153.

Kerr, Richard A. "Precision of Global Positioning Increases." *Science* Vol. 236 (June 26, 1987): 1625.

McElroy, John H. *Space Science and Applications* IEEE Press, 1986.

Weisburd, Stefi. "'Seeing' Continents Drift." *Science News* Vol. 128 (December 21, 1985): 388.

ASSESSING AGRICULTURE

Brown, Lester R. "World Population Growth, Soil Erosion, and Food Security." *Science* Vol. 214 (November 27, 1981): 995-1001.

Carey, William D. "How in the World Are We?" *Science* Vol. 232 (May 30, 1986): 1073.

Gibbons, Boyd. "Do We Treat Our Soil Like Dirt?" *National Geographic* Vol. 166 (September 1984): 353-388.

Larson, W. E., F. J. Pierce, and R. H. Dowdy. "The Threat of Soil Erosion to Long-term Crop Production." *Science* Vol. 219 (February 4, 1983): 458-464.

Mohler, Robert R. J., et al. "Monitoring vegetation of drought environments." *BioScience* Vol. 36 (July/August 1986): 478-483.

Paul, Charles K., and Adolfo C. Mascarenhas. "Remote Sensing in Development." *Science* Vol. 214 (October 9, 1981): 139-145.

Peterson, Ivars. "Keeping the topsoil down on the farm." *Science News* Vol. 132 (December 5, 1987): 357-358.

Raloff, Janet. "Salt of the Earth." *Science News* Vol. 126 (November 10, 1984): 298-301.

_____. "Soil Losses Eroding Food Security." *Science News* Vol. 126 (October 6, 1984): 212.

Tucker, C. J., et al. "Relationship between atmospheric CO_2 variations and satellite-derived vegetation index." *Nature* Vol. 319 (January 16, 1986): 195-198.

Walsh, John. "Famine Early Warning Closer to Reality." *Science* Vol. 233 (September 12, 1986): 1145-1146.

Warrick, R. A. "Photosynthesis seen from above." *Nature* Vol. 319 (January 16, 1986): 181.

"World Trends & Forecasts." *The Futurist* (September/October 1986): 44-47.

DEPICTING DISASTERS

Anderson, Harry. "A Deadly Wall of Water." *Newsweek* (June 10, 1985): 56-57.

Fisher, Arthur. "Eastern Earthquake" *Popular Science* Vol. 130 (May 1987): 9-12.

_____. "Predicting Volcanoes." *Popular Science* Vol. 231 (August 1987): 10-12.

Hafemeister, David, Joseph J. Romm, and Kosta Tsipis. "The Verification of Compliance with Arms-Control Agreements." *Scientific American* Vol. 252 (March 1985): 39-45.

Heppenheimer, T. A. "Finally a saturation attack on earthquake Prediction." *Popular Science* Vol. 229 (August 1986): 54-58.

Kerr, Richard A. "Earthquakes Are Giving Little Warning." *Science* Vol. 235 (January 9, 1987): 165-166.

Revkin, Andrew C. "Waiting for the Big One." *Discover* Vol. 9 (January 1988): 73-74.

Robok, Alan. "The dust cloud of the century." *Nature* Vol. 301 (February 3, 1983): 373-374.

Tsipis, Costa. "Arms Control Pacts Can Be Verified." *Discover* Vol. 8 (April 1987): 79-93.

Weisburd, Stefi. "Radio waves signal earthquakes." *Science News* Vol. 130 (December 20 & 27, 1986): 399.

Wesson, Robert L., and Robert E. Wallace. "Predicting the Next Great Earthquake in California." *Scientific American* Vol. 252 (February 1985): 35-43.

Index

Index

185

Other Bestsellers From TAB

☐ **SUPERCONDUCTIVITY: Experimenting in a New Technology—Dave Prochnow**

The latest addition to TAB BOOKS' Advanced Technology Series . . . Now, you have a chance to expand your own knowledge and understanding of superconductivity and to begin performing your working superconductivity experiments. Written for the advanced experimenter, this book includes specific equations and references to the chemistry, thermodynamics, and quantum mechanics that explain this phenomenon. 160 pp., Illustrated.

Paper $17.95   **Hard $22.95**
Book No. 3132

☐ **SUPERCONDUCTIVITY—The Threshold of a New Technology—Jonathan L. Mayo**

Superconductivity is generating an excitement not seen in the scientific world for decades! Experts are predicting advances in state-of-the-art technology that will make most exisiting electrical and electronic technologies obsolete! This book is one of the most complete and thorough introductions to a multifaceted phenomenon that covers the full spectrum of superconductivity and superconductive technology. 160 pp., 58 illus.

Paper $10.95   **Hard $12.95**
Book No. 3022

☐ **LASERS—The Light Fantastic—2nd Edition —Clayton L Hallmark and Delton T. Horn**

Gain insight into all the various ways lasers are used today . . . in communications, in radar, as gyroscopes, in industry, and in commerce. Plus, more emphasis is placed on laser applications for electronics hobbyists and general science enthusiasts. If you want to experiment with lasers, you will find the guidance you need here—including safety techniques, a complete glossary of technical terms, actual schematics, and information on obtaining the necessary materials. 280 pp., 129 illus.

Paper $15.95   **Hard $19.95**
Book No. 2905

☐ **UNDERSTANDING MAGNETISM: Magnets, Electromagnets and Superconducting Magnets —Robert Wood**

Explore the mysteries of magnetic and electromagnetic phenomena. This book bridges the information gap between children's books on magnets and the physicist's technical manuals. Written in an easy-to-follow manner, *Understanding Magnetism*, examines the world of magnetic phenomena and its relationship to electricity. Thirteen illustrated experiments are provided to give you hands-on understanding of magnetic fields. 176 pp., 138 illus.

Paper $13.95   **Hard $17.95**
Book No. 2772

☐ **GALLIUM ARSENIDE IC TECHNOLOGY: Principles and Practice—Neil Sclater**

Now you can explore this decade's most exciting breakthrough in integrated circuit fabrication technology! With this book, Neil Sclater offers you the key to being on the cutting edge of the fastest growing field in semiconductor technology. Here is an excellent, nonmathematical overview of gallium arsenide (GaAs) ICs: how they are manufactured and packaged, what benefits they provide, and what the future holds for these innovative devices. 272 pp., 153 Illus.

Paper $21.95   **Hard $26.95**
Book No. 3089

☐ **SCIENCE FAIR: Developing a Successful and Fun Project—Maxine Hern Iritz, Photographs by A. Frank Iritz**

Here's all the step-by-step guidance parents and teachers need to help students complete prize-quality science fair projects! This book provides easy-to-follow advice on every step of science fair project preparation from choosing a topic and defining the problem to setting up and conducting the experiment, drawing conclusions, and setting up the fair display. 96 pp., 83 illus., 8 1/2″ × 11″.

Paper $11.95   **Hard $14.95**
Book No. 2936

☐ **PUZZLES, PARADOXES AND BRAIN TEASERS —Stan Gibilisco**

Explore the loopholes in mathematical logic! This is a clear, concise, well-written exploration of the mysteries of the universe. It is an intriguing look at those exceptions that are as frustrating as they are amusing! The author's approach is entertaining, enlightening, and easy to understand. Although the topics are of a mathematical nature, the discussions are nontechnical. 122 pp., 83 illus.

Paper $11.95   **Hard $14.95**
Book No. 2895

☐ **BUILD YOUR OWN WORKING FIBEROPTIC, INFRARED AND LASER SPACE-AGE PROJECTS—Robert E. Iannini**

Here are plans for a variety of useful electronic and scientific devices, including a high sensitivity laser light detector and a high voltage laboratory generator (useful in all sorts of laser, plasma ion, and particle applications as well as for lighting displays and special effects). And that's just the beginning of the exciting space age technology that you'll be able to put to work! 288 pp., 198 illus.

Paper $18.95   **Hard $24.95**
Book No. 2724

Other Bestsellers From TAB

Help Us Help You

So that we can better provide you with the practical information you need, please take a moment to complete and return this card.

1. I am interested in books on the following subjects:

☐ architecture & design
☐ automotive
☐ aviation
☐ business & finance
☐ computer, mini & mainframe
☐ computer, micros
☐ other_____

☐ electronics
☐ engineering
☐ hobbies & crafts
☐ how-to, do-it-yourself
☐ military history
☐ nautical

2. I own/use a computer:

☐ Apple/Macintosh_____
☐ Commodore_____
☐ IBM_____
☐ Other_____

3. This card came from TAB book (no. or title):

_____

4. I purchase books from/by:

☐ general bookstores
☐ technical bookstores
☐ college bookstores
☐ mail

☐ telephone
☐ electronic mail
☐ hobby stores
☐ art materials stores

Comments _____

_____

_____

Name _____

Address _____

City _____

State/Zip _____

TAB BOOKS Inc.